Environmental Footprints and Eco-design of Products and Processes

Series editor

Subramanian Senthilkannan Muthu, SGS Hong Kong Limited,
Hong Kong, Hong Kong SAR

More information about this series at http://www.springer.com/series/13340

Subramanian Senthilkannan Muthu
Miguel Angel Gardetti
Editors

Sustainable Fibres for Fashion Industry

Volume 2

 Springer

Editors
Subramanian Senthilkannan Muthu
Environmental Services Manager-Asia
SGS Hong Kong Limited
Hong Kong
Hong Kong SAR

Miguel Angel Gardetti
Center for Studies on Sustainable Luxury
Buenos Aires
Argentina

ISSN 2345-7651   ISSN 2345-766X   (electronic)
Environmental Footprints and Eco-design of Products and Processes
ISBN 978-981-10-9195-7   ISBN 978-981-10-0566-4   (eBook)
DOI 10.1007/978-981-10-0566-4

Printed on acid-free paper

This Springer imprint is published by Springer Nature
The registered company is Springer Science+Business Media Singapore Pte Ltd.

Preface

The need for sustainable fibre sources is inevitable—recognizing this is the first step toward attaining sustainable development in the entire textiles and clothing industry. Hence we decided to put together dedicated volumes with the express purpose of dealing with this important subject. This is the second volume on sustainable fibres for the fashion industry—is a continuation of the first. This volume comprises six well-written chapters that deal with sustainable fibres for the fashion industry.

This volume begins with a chapter written by *Geetha Dissanayake and Srimala Perera*, titled **New Approaches to Sustainable Fibres**. The chapter chiefly deals with the basic concept of "design to compost" the purpose of which is to address the challenges facing sustainable fashion. A systematic review on current trends in nature inspired the authors to investigate sustainable textile fibres. Detailed discussions pertaining to potential research avenues for compostable textile fibres are also dealt with.

The second chapter titled **Consumer Perceptions of Fibres with Respect to Luxury and Sustainability: An Exploratory Study**, authored by *Cathy A. Rusinko and Marie-Eve Faust*, is a detailed study of consumer perceptions of fibres with respect to luxury and sustainability. This study comprises a survey, consisting of theory-grounded questions put to a sample of young consumers. The findings show that young consumers have a tendency to perceive most fibres (with the exception of two fibers) as either luxurious or sustainable—but not both.

The third chapter on **Sustainable Natural Fibres from Animals, Plants and Agroindustrial Wastes—An Overview**, written by *Shahid-ul-Islam and Faqeer Mohammad*, highlights sources and important characteristics of various sustainable natural fibres derived from various natural sources: namely wool, cotton, ramie (a flowering plant in the nettle family) and jute. Discussions related to the production of cellulose and protein-based natural fibres from agricultural wastes are also included.

The fourth chapter titled **Sabai Grass: Possibility of Becoming a Potential Textile**, authored by *Asimananda Khandual and Sanjay Sahu*, examines the possibility of utilizing sabai grass (*Eulaliapsis binata*) as a sustainable textile fibre source for different textile applications. The chapter presents the history and socioeconomic importance, chemical constitutions, fibre properties, recent research and potential application alternatives for sabai grass fibre.

The fifth chapter titled **Potential of Ligno-cellulosic and Protein Fibres in Sustainable Fashion**, written by *Kartick K. Samanta, S. Basak, S.K. Chattopadhyay and P. Chowdhury*, presents an overall picture of various lignocellulosic and protein fibres for various textile applications in the fashion industry. The authors take a complete journey around various important aspects (production, physical and chemical properties, applications) of these fibres and the fabrics made from them.

The sixth chapter titled **Milkweed—A Potential Sustainable Natural Fibre Crop**, authored by *T. Karthik and R. Murugan*, revolves around milkweed fibres and fabrics made from them. The chapter presents a good deal of information pertaining to the history of the milkweed plant and fibers such as fiber morphology and characteristics, spinnability, fabric properties and potential application in clothing, lightweight composites, oil sorption, and thermal and acoustic insulation.

We are confident that readers of the two volumes of *Sustainable Fibres for Fashion Industry* will get a lot of very useful information about the various sustainable fibres employed in the fashion sector. We would like to mark our sincere thanks to all the authors who have contributed the six chapt. in this second volume for their time and effort.

Subramanian Senthilkannan Muthu
Miguel Angel Gardetti

Contents

New Approaches to Sustainable Fibres

Geetha Dissanayake and Srimala Perera

Abstract Sustainability is becoming a prerequisite to the future of the fashion industry. A sustainable vision of the fashion industry relies on fibre, material and product-based innovations. Sustainable fibres not only contribute to solving the environmental burden, but also are becoming a new trend in the fashion market. Moreover, the focus of the fashion industry today is moving beyond the design aspects of products into functional and therapeutic features. While the term "sustainable fibre" still accentuates organic cotton or recycled polyester, most other innovations addressing sustainability together with desirable features are stuck in the niche. The journey towards sustainable fibers is seemingly challenging with many failures. However, exciting innovations are emerging fast, challenging the traditional ways of producing fibres. One such way is emulating nature (biomimicking) in fibre-to-fabric production. Biomimicking research in textile production is a rapidly growing area and its true potential in the development of entirely sustainable fibres has yet to be discovered through interdisciplinary research with an understanding of the holistic approach of nature in its formation of organisms. Nature provides excellent examples of complex functional systems, which are created through entirely sustainable processes. Most importantly, those natural systems leave no trace of waste at the end of their lives. Keeping nature's way in mind, this chapter pushes the concept of design to compost to address the challenges facing sustainable fashion. The chapter also reviews some examples of current trends in nature that inspire sustainable textile fibers and discusses potential research avenues for compostable textile fibers, which need to take on board the zero-waste approach of nature.

G. Dissanayake (✉)
Department of Textile and Clothing Technology, University of Moratuwa,
Moratuwa, Sri Lanka
e-mail: geethadis@uom.lk

S. Perera
Institute of Technology, University of Moratuwa, Moratuwa, Sri Lanka

© Springer Science+Business Media Singapore 2016
S.S. Muthu and M.A. Gardetti (eds.), *Sustainable Fibres for Fashion Industry*,
Environmental Footprints and Eco-design of Products and Processes,
DOI 10.1007/978-981-10-0566-4_1

1 Introduction

Mass production and fast supply chains have become the buzzwords of every business in today's competitive world. In the textile industry, this had led to increased production of fibres, yarns and fabrics at an incredible rate. As a result of the variety of textiles available today the fashion industry is rapidly moving from traditional four-seasonal fashion collections to a fast fashion business that offers new collections almost every month. Fast fashion produces disposable clothing by mimicking luxury fashion brands [16], where the introduction and obsolescence of new trends occur rapidly. The fast replication of fashion and mass-scale production significantly reduce the price and quality of the final product. The availability of cheap clothes gradually changes the mindsets of people from thrift to extravagance. Landfills are starting to overflow with clothes because businesses and consumers overlooked what to do when those clothes are no longer needed.

More than six billion people inhabit our planet today (as of 2016) and they produce millions of tons of waste each day. It is almost unimaginable that textile waste is forecast to be the fastest growing waste category between 2005 and 2020 [7]. Most textile waste requires years to decay and takes up significant land space, directly threatening the wellbeing of organisms on the planet. Recently, the invention of fast-response systems promoting overabundance and disposability as well as increasing the demand for fast fashion has made the situation much worse [13, 32]. Consumers have started purchasing fashion clothing and throwing them away at a profligate rate and waste has accumulated. Most waste ends up in landfills without any kind of treatment. These materials hold compounds that release toxic chemicals or take unpredictably long times to break down. Only a fraction, a very small percentage, is recycled or remanufactured, although that is insufficient to offset the damage being done to the environment by the textile and fashion industry. This chapter looks at fashion production from a new perspective and argues that "compostable fashion", which is made from compostable fibres and leaves no waste when discarded, would be one of the best ways to enhance the sustainability of the fashion industry.

2 Challenge of Sustainable Fashion

Usually, fashion and sustainability are often contradictory [8, 21] and the challenge for the fashion industry today is to find the best possible ways to merge them. While there is no universal definition of "sustainable fashion", many people in the fashion industry lok at it from the perspective of their own positions in the process. At the beginning of this process, a textile producer might look at it from the perspective of energy consumption, water consumption, toxic chemicals in the process and working conditions in the mills. Next, a concerned fashion designer would identify it as promoting an eco-label. Then, for a garment manufacturer, it could be fair and safe working conditions for his production team. Finally, for the customer, it could be enduring usage, recyclability, etc. Without any doubt, sustainable fashion is all of these and more and demands a holistic viewpoint, a closed-loop production system brought about by a cradle-to-cradle approach [11, 20]. As consumers are increasingly becoming aware of the environmental issues associated with fashion and demanding sustainable choices, retailers and manufacturers need to think beyond profits. While it is still unclear how the textile and fashion industry could turn into a completely sustainable business, reducing the sustainability impacts of fibres is one of the key priorities.

Textile fashion production involves one of the most complicated supply chains in the manufacturing sector [13], and environmental issues arise at all stages of production. Major environmental impacts arise from the use of energy, water, toxic chemicals and depletion of natural resources [1, 19]. The textile and fashion supply chain usually starts with the production of natural or synthetic fibres. Natural fibres are obtained from plants or animals, and man-made fibres are produced by using petroleum-based chemicals. The production of both natural and synthetic fibres has a significant impact on the environment. Cotton, one of the most versatile natural fibres used in textile manufacturing, is a heavily water- and pesticide-dependent crop. While conventional cotton is often criticized for its use of water, land, pesticides and resultant adverse impacts on the environment during farming, organic cotton cultivation is associated with even higher land use due to reduced yields [34]. No matter whether cotton is conventional or organic, this crop always leaves a high water footprint.

Synthetic fibres such as polyester use comparatively less water in their production than cotton; however, energy requirements for the production of synthetic fibres create a higher carbon footprint. Major concerns associated with synthetic fibres are the consumption of non-renewable resources to produce fibres which are often non-biodegradable. While cotton and polyester fibres dominate the industry today, emerging natural fibres such as hemp, jute, and banana are struggling to gain market access [34]. With increasing world population, rising oil prices (although they are currently in freefall) and changing consumption patterns, it is impossible to consume natural resources to fulfil fibre demand without creating further damage to the environment. Nevertheless, finding alternative sources for fibre production is a real challenge. There is a trend towards producing compostable fibres [4] because environmentally friendly fibres can make a difference in reducing the environmental footprint of the textile and fashion industry.

Recently, the advent of advanced technologies has paved the way for researchers to look at fashion from some other perspectives. They have started to incorporate different functionalities to textiles to extend the useful life of clothing. Fashionable garments with different functionalities—stain repellent, antimicrobial and wrinkle resistant—have rapidly emerged. To add these functionalities to fabrics many chemicals are used at different stages in the production of fibre, yarn or textiles. Although these functionalities lengthen the useful life of textiles and garments the fabric treatment methodologies used in the process add many new problems to the

Fig. 1 Material recirculation loop for fashion clothing (*Source* Authors)

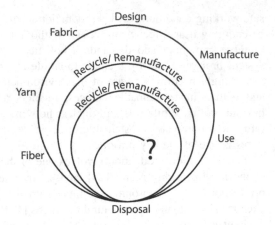

sustainability issue such as loss of desirable properties and uncertainty about the toxicity of fabrics [5, 25].

The demand for textile fibres has doubled over the last few decades [34], and this demand is heavily influenced by fast fashion. This means the rate of use exceeds the rate at which natural resources can be renewed. We cannot grow cotton or extract fossil fuel at the rate demanded by fast fashion. Currently, resources are depleting and waste is being generated at a faster rate than ever before, but material recirculation is being highly promoted through the recycling or remanufacturing of fashion waste. However, as shown in Fig. 1, materials cannot be recirculated endlessly. How many times can fashion clothing be recycled or remanufactured? There is no definitive answer yet, but extending the product's life more than once is really challenging. Moreover, recycling is often known as "downcycling", where the quality and properties of original products are downgraded. Producers use energy and chemicals in the recycling process, creating a greater environmental burden, yet there is no guarantee that recycled fashion would remain longer in consumers' wardrobes than ordinary fashion clothing. As long as fashion clothing is considered disposable, fashion-minded consumers will buy and discard it as usual. Therefore, converting fashion waste back to fashion clothing by recycling or remanufacturing would only address short-term issues, such as diversion of waste from landfills—a stopgap until a definite clear route to an entirely sustainable world of textile and fashion is found.

Today, the fashion industry bears the responsibility to start new product development with a view to changing the entire process, not only to maximize *end profit*, but also to minimize *end waste*. To reduce the massive amount of waste generation, one clear option is to slow down the rate of consumption and disposal habits, allowing sufficient time for nature to renew and automatically balance the ecosystem. However, this requires a major shift in the mindset of the consumer, which is almost impossible in today's modern society as "consumption has many meanings for the consumer" [22]. Because there is little light at the end of the tunnel in changing consumer behaviour the next best option is to reduce the environmental burden caused by overconsumption and subsequent waste. Imagine waste being naturally composted, without releasing any harmful substances to the environment. In this context, the focus need not be essentially on changing fast fashion to slow fashion, but on changing "any fashion" to "fast compost". Nature gives us plenty of inspirations for this. Consider a mature cotton seed—it falls down from the cotton tree, finds its way into the soil and later sprouts under the right conditions. It will create a new generation or will just compost. It is a complex production system, but there is no waste—it simply returns natural elements to our ecosystem. However, if the cotton seed is used to create a simple cotton garment, it undergoes numerous processes in the fabric mill and finally ends up as a complex product. At this stage it is really difficult to find its natural biodegradable root, let alone methods to recycle it. Thus, the challenge is obvious. The fashion process should produce goods that leave no trace once discarded, in line with nature's design for composting.

3 Nature-Inspired Sustainable Fibres: Biomimicry

Sustainability is best described by nature. If you observe a leaf that has fallen from a tree after a few days, you will see that it has decomposed eventually leaving no trace. Having a surface that is highly complex and multifaceted—water repellent, efficient water dragging, insect repellent, thermally insulated—the leaf finally buries itself into the soil and disappears leaving no trace. As Benyus [2] explained, "nature is a genius of design", and we add to that, "*nature is a genius to the core*". This connotation works best because from start to end, nature harmonizes sustainability by creating all the elements of nature and making the best use of resources without any waste. Apart from being highly functional, most biological structures in nature have the ability to produce a range of aromas and to manipulate light into a vibrant range of colours. With the help of advanced technology, especially nanotechnology, scientists ae able to look deep into different biological structures to adopt a new perspective to functional, aromatic, and colourful fibres in textiles [3, 6, 24, 29, 37]. This concept of emulating nature's way is called "biomimicry".

Scientists are successfully incorporating knowledge of biological structures to impart many functionalities to textiles—such as being water repellent, stain repellent, self-repairing, thermo-insulating and antimicrobial. However, the resultant fibres and fabrics are little more than copies of nature's concepts, and a question mark still lingers over the environmental impact. Most of these functionalities are added to textiles by introducing more chemicals to the base material. However, this type of biomimicry can mask the desirable performance and aesthetic properties of the base material. If not assessed carefully during synthesis, it could end up adding more toxicity to the environment. The obvious shortcoming of linear progression of science into technology and then to the consumer is also reflected here in the adoption of biomimicry for textiles. Instead, we argue that biomimicry in sustainable fibres and fabrics needs to take an environmental empathetic approach. A number of such alternative approaches are discussed below.

Fibre is the basic element of textiles. In nature, fibers are found in many forms: from nanoscale silk fibres produced in the cribellar glands of spiders to micro-scale cellulose fibers in wood [28, 35]. Most of these fibres fulfil a multitude of elementary functions of respective lifeforms. A closer look at the natural fibres found in animals and plants reveals that they are made up of polypeptides (proteins) and polysaccharides (cellulose and chitin) which are biodegradable making it easier for the lifecycles of these organisms to end up leaving no undesirable trace. However, the question is whether we take the maximum benefit of these properties into our textiles. The availability of a variety of natural and synthetic fibres today often leads to using fibre blends in the production of textiles. Moreover, some other functional properties are added to the textiles by mixing and coating many different materials. As a result the original identity of the fibre is hidden or lost in the process, which makes the product impossible or very difficult to recycle or compost [10].

The challenge of design to compost in sustainable fashion can be addressed by giving more emphasis to this non-blended material concept in nature. The development of bio-inspired polymers is one of the most promising pathways towards this concept. Researchers are already trying to mimic spider and insect silk fibres and marine mussel byssus threads as models for novel fibre manufacturing [15, 31]. These protein-based biopolymers are found to have exceptional mechanical properties and, most importantly, they are being spun at ambient temperatures and pressures with water as the solvent [36]. Recently, methods to spin artificial spider silk have gained much attention. In one such method, silk-producing genes obtained from spiders were inserted into mammalian cells to obtain recombinant spider silk proteins. An aqueous solution of these proteins was then used to spin silk monofilaments [17]. Similarly, recombinant spider silk proteins produced using bacteria and yeast cell cultures doped with chemically developed genes have been reported to have been used in solvent-based spinning of silk [23].

Plants and trees also provide other excellent examples of bio-inspired fibrous structures. For example, Lenpur is a sustainable fibre introduced to the textile market with design to compost in mind. This fibre is made from white pine tree clippings. Only certain parts of the tree are collected in the clipping stage to produce a wood pulp, which is then treated in a rayon process to create Lenpur fibre. The fabric imparts excellent comfort and hand (feel) properties due to its softness, its absorption capacity and ability to release moisture (due to fibre morphology). Laboratory tests have shown that the absorption capacity of Lenpur fibre is 35 % higher than that of cotton and almost double that of viscose [26]. Further, the fibre's ability to sustain a higher thermal range gives it an advantage over other cellulosic fibres as it keeps the wearer cooler in the summer and warmer in the winter. Lenpur fabric is used for performance garments and undergarments due to its exceptional human-friendly properties.

Lenpur fits well with the closed-loop concept not only because it is compostable but also because it involves a sustainable production process. The clippings taken from the white pine tree, which are used to produce the fabric, are only those that do not contribute to the continued life of the tree. In fact, no forests are harmed and the bark and branches are gathered through regular pruning. Therefore, the tree is left to continue its life. Most importantly, Lenpur undergoes a unique production process, which is due to the self-mercerizing (lustre-giving) quality and high dye intake of the fibre. Thus, fewer chemicals are used in this process compared with the standard coloration process. This fabric has already gained popularity among eco-designers who are specialized in using fabrics that are made from sustainable products in their fashion lines.

Freitag, a Swiss-based company, also produces a compostable fabric—called "F-abric". The fabric is used to manufacture workwear for its employees and designed to be composted at the end of the product's life. By using linen, hemp and modal fibres which do not require excessive amounts of water to grow, the company produces cotton-free denims, work shirts, pants and T-shirts. The fabric will biodegrade completely in a few months much like garden compost, leaving no footprint [14]. This concept is nature friendly and very useful in reducing adverse environmental impacts caused by massive waste flows. As a compostable fabric, it must not release any harmful substances to the environment or leave any residue, so the full lifecycle of the fibre needs to be carefully designed and developed—from cradle to grave. However, design to compost does not add any functional or economic value to the product and, therefore, is hard to promote among profit-minded fashion manufacturers unless the responsibility for waste management is given to manufacturers and retailers by legislation.

Another exciting inspiration we can acquire from nature is its production of brilliant colours. Indeed, colour is an essential element of fashion textiles. At present, there are but a few sustainable, natural textile dyeing practices which are mostly based on pigments extracted from parts of trees. However, scaling up natural dye production and the durability of these dyes on fabrics still poses challenges to the industry. The answer lies in nature, that provides some models for us to follow. Nature often equips its creatures with genes or complex photonic structures that can produce a variety of colours in stark contrast to the pigment colours we often use in the conventional colouration process.

Cotton dominates the natural fibre category of textile raw materials. For some while now, scientists have been trying to modify cotton genes to harvest new varieties of colourful cotton. Many successful cultivars have been grown for years in South and Central America, Russia, India and Israel. Currently, shades of brown and green are popular, yet efforts are being made to achieve a diverse colour range by transferring genes from flowering plants [12]. Already coloured fibres avoid the chemical-intensive ordinary dyeing process of fabrics and would become effective, economical and sustainable if a range of colours could be obtained.

The demand for fluorescent colors in today's fashion industry is high. However, many toxic chemicals are utilized in the fluorescent fabric–dyeing process. Recently, scientists have been able to create naturally coloured silk cocoons by genetically modifying the DNA structures of silkworms. Although not commercialized yet, a group of Japanese researchers have genetically engineered silkworms to produce fluorescent skeins [18], which is undeniably a wonder considering the

Fig. 2 Wedding dress
produced using coloured
fluorescent silks (*Source* [18])

toxic waste normally generated in the standard process. This Japanese glow-in-the-dark silk comes in shades of red, orange and green. The Japanese wedding dress designer Yumi Katsura has already incorporated glowing silks into daywear such as suits and ties, and has designed gowns that glow in the dark (as shown in Fig. 2). These glowing silk–producing worms are a result of genome mutation done by scientists using DNA sequences that produce a range of foreign fluorescent proteins. Though we find fluorescent colours fashionably inspiring, nature has granted it to some creatures as a survival technique. Thus, scientists have borrowed fluorescent proteins from a few of these creatures: glowing red from *Discosoma* corals, orange from *Fungia concinna* coral and green from jellyfish. In addition to this peculiar silk-colouration method the processing steps involved in converting these glowing cocoons into silk fabric are also slightly different from those of normal silk [18]. In this revolutionary technique which eliminates the colouration process, no hazardous waste is emitted and no water is wasted.

Some of the beautiful and vibrant colours we see in nature are not the result of pigments, but of the reflection or diffraction interaction of light from structures, which is known as "structural colours". Structural colours are created by optical interference of light reflected from nanoscale structures [9, 27]. The colours of butterflies, peacock feathers, beetles and skins of squid are just a few examples of structural colours [29]. These structures suggest a new perspective on colouration. To mimic these wonderful natural creations in fibres and fabrics, we need to understand the structures and how they create colours, which would point up new ways of colouring the fibres and fabrics. Tejin, a Japanese company, has successfully mimicked the nanostructure of butterfly wings to produce a structurally coloured polyester fibre called "Morphotex"; the colour is created by varying the structure and thickness of fibres [30]. Even though production has been discontinued, this invention revealed the possibility of creating colours for fabrics using structures. If fabrics could be structurally coloured the manufacturing process could eliminate the colouration process of fabrics. Moreover, this type of innovation would support the concept of design to compost if fibres are extracted from natural sources and no artificial colouration process is being used—as such the final product would be a completely natural element.

The use of biomimicry in the development of sustainable fibres is a promising development towards sustainability in the fashion industry. It is an effort to unite economy and technology with the environment. Designers should investigate nature on how to mimic natural innovations and, thereby, bring sustainable solutions to the textile and fashion industry. Regardless of the sustainability potential of biomimicry in textiles, it does not appear as yet to be an explicit innovation target of the textile industry.

4 Recommendations for Further Research

The law of flotation was not discovered by contemplating the sinking of things, but by contemplating the floating of things which floated naturally, and then intelligently asking why they did so [33].

As Thomas Troward pointed out, our ancestors observed that wood floats in water, which led to building ships from wood. They also noticed that iron sinks. Therefore, iron was not used for shipbuilding. However, when the law of flotation was discovered, people realized that anything could float as long as it is lighter than the mass of the liquid it displaces, and this discovery led to iron dominating the shipbuilding industry.

The key idea in Troward's quote is of paramount importance to addressing the key challenges facing sustainable fashion. It is now very clear that a paradigm shift in how sustainable fashion is perceived needs a positive response. One can argue that this shift is moving away from the culture of fast fashion to slow fashion. Moreover, one could focus on discovering more recycling or remanufacturing roots to slow down the generation of waste. However, such arguments are clearly short sighted and, as Troward pointed out, could not be considered solutions to challenges discovered by *contemplating the sinking of things*.

Globalization drives the culture of fast fashion. Currently, the drive used to create an empathetic view of fast fashion and to promote the idea of slow fashion is little different. However, this gradual shift requires time, measured not in months or years but in decades or generations. As discussed earlier, recycling and remanufacturing which do not equate with models in nature always lead to a question mark (Fig. 1). Therefore, a different approach to address the challenges facing sustainable fashion is absolutely necessary. We suggest the shift should be directed towards nature. We must try and discover the mechanisms that drive nature's incessant creation of organisms without piling up mountains of waste. Researchers have already begun the study of biodegradation, mineralization and biomass formation, which is nature's way of creating zero waste [4]. Discovery of the laws of zero waste in nature could then be mimicked in the production of fast-compostable textile fibres.

Further, nature creates complex and toxic-free functionalities—such as water repellence—using surface structures with the help of many different cell types, cell surface structures and cell shapes. In this context, we suggest that the focus should be on mimicking these structures physically in fibres and fabric structures rather than applying toxic chemicals to achieve such functionalities. Mimicking the concept of structural colours in fabrics is an exciting discovery in this area; however, many more innovations are awaited. Finally, it is the responsibility of textile engineers to look at ways of mimicking natural wonders and incorporate them in textile fibers—to innovate the next generation of fast-compostable fibers.

5 Conclusion

The discussion about fashion and sustainability is gaining momentum, and sustainable fibres occupy centre stage of the discussion. As cotton and oil prices are increasing (although the latter are in freefall currently, as of 2016) and cotton and petroleum-based fibres are creating harm to the environment, more environmentally responsible alternatives need to be found. However, sustainable fibres will remain in the niche unless there is a commitment to getting sustainability, functionality and upscaling to work together. Survival of the fashion industry is questionable in its current status unless we slow down production, allow sufficient time for resources to renew or increase sustainable fibre production to fulfil ever increasing market demand without creating further damage to the environment.

The demand for sustainable fibres has grown significantly in recent years, indicating strong interest among producers, retailers and consumers. Increasing awareness of the environmental implications of production, consumption and disposal of conventional fibres has fuelled the demand for eco-friendly fibres. The future of the sustainable fibre industry looks bright and growth is expected to be stimulated due to the increasing commitment of manufacturers and retailers; however, the real challenge is in the supply side of sustainable fibres. Some fibres have been found to be sustainable, yet have adverse impacts on the environment during processing, adding functional properties or when disposing after use. Indeed, it is difficult to decide whether one fibre is more sustainable than the next; a full lifecycle analysis is required to evaluate the environmental impact during the life of a product. A sustainable fibre needs to be environmentally friendly throughout its production phase and expect to decay naturally after being released to the environment. This chapter suggests that the next generation of textile fibers should be developed emphasizing design to compost, where fashion, created using those fibres, becomes little more than a part of nature when disposed of.

A close study of nature can open up many avenues toward design to compost. Understanding natural degradation processes is suggested as key to developing sustainable textile fibers. Furthermore, a close study of the physical structures in nature with respect to their complex functionalities—such as being hydrophobic, thermo-regulating or self-healing—is highly encouraged in an attempt to mimic nature and create sustainable fibres without using toxic chemicals. In fact, animal and plant fibre structures and assemblies readily provide exemplary ideas for use in the design-to-compost approach.

References

1. Allwood J, Laursen SE, de Rodríguez CM, Bocken NMP (2006) Well Dressed? The present and future sustainability of clothing and textiles in the United Kingdom. University of Cambridge Institute for Manufacturing
2. Benyus JM (1997) Biomimicry: innovation inspired by nature. HarperCollins
3. Bhushan B (2009) Biomimetics: lessons from nature—an overview. Philos Trans R Soc A 367:1445–1486
4. Blackburn R (2005) *Biodegradable and Sustainable Fibres.* Woodhead Publishing Series in Textiles.
5. Crosera M, Bovenzi M, Maina G, Adami G, Zanette C, Florio C, Larese FF (2009) Nanoparticle dermal absorption and toxicity: a review of the literature. Int. Achieves Occup. Health 82:1043–1055
6. Dawson C, Vincent JFV, Jeronimidis G, Rice G, Forshaw P (1999) Heat transfer through penguin feathers. J Theor Biol 199:291–295
7. Defra (2006) Modelling the impact of lifestyle changes on household waste arisings. http://randd.defra.gov.uk/Document.aspx?Document=WR0107_8322_FRA.pdf
8. Dissanayake G, Sinha P (2012) Sustainable waste management strategies in the fashion industry sector. Int J Environ Sustain 8(1):77–90
9. Dodson B (2013) Fruit-inspired fibers change color when stretched. http://www.gizmag.com/biomimetic-fiber-color-changing-stretch/26024/
10. Eadie L, Gosh TK (2011) Biomimicry in textiles: past, present and potential. an overview. J R Soc Interface 8:761–775
11. Eagan GJ (2010) The sustainable fashion blueprint: completing the loop. Published Master's Thesis. University of Buckingham, UK
12. Fibre2fashion (2011) Indian scientists grow eco-friendly coloured cotton. http://www.fibre2fashion.com/news/textile-news/newsdetails.aspx?news_id=105720
13. Fletcher K (2008) Sustainable fashion and textile: design journeys. Earthscan. London, Washington, DC
14. Freitag (2015) FREITAG F-ABRIC. http://www.freitag.ch/fabric
15. Gruwald I, Rischka K, Kast SM, Scheibel T, Bargel H (2009) Mimicking iopolymers on a molecular scale:nano(bio)technology based on engineered proteins. Philos Trans R Soc 367:1727–1747
16. Joy A, Sherry JF Jr, Venkatesh A, Wang J, Chan R (2012) Fast fashion, sustainability and the ethical appeal of luxury brands. Fashion Theory 16(3):273–296
17. Lazaris A, Arcidiacono S, Huang Y, Zhou J, Duguay F, Chretien N, Welsh EA, Soares JW, Karatzas CN (2002) Spider silk fibers spun from soluble recombinant silk produced in mammalian cells. Nature 295:472–476
18. Lizuka T, Sezutsu H, Tatematsu K, Kobayashi I, Yonemura N, Uchino K, Nakajima K, Kojima K, Takabayashi C, Machii H, Yamada K, Kurihara H, Asakura T, Nakazawa Y, Miyawaki A, Karasawa S, Kobayashi H, Yamaguchi J, Kuwabara N, Nakamura T, Yoshii K, Tamura T (2013) Colored fluorescent silk made by transgenic silkworms. Adv. Funct. Mater. 23(42):5232–5239
19. Lorek S, Lucas R (2003) Towards sustainable market strategies. Wuppertal Papers
20. McDonough W, Braungart M (2002) Cradle to cradle. remaking the way we make things. Vintage
21. Nielsen (2013) Sustainability and fashion: the case of bamboo. http://skemman.is/stream/get/.../BA_-_2013_-_Bethina_Elverdam_Nielsen.pdf

22. Niinimäki K (2015) Consumer behaviour in the fashion field. In: Muthu SS (ed) Handbook of sustainable apparel production. CRC Press
23. O'Brien JP, Fahnestock SR, Termonia Y, Gardner KH (1998) Nylons from nature: synthetic analogs to spider silk. Adv Mater 10:1185–1195
24. Parker AR (1998) The diversity and implications of animal structural colours. J Exp Biol 201:2343–2347
25. Perera S, Bhushan B, Bandara R, Rajapaksha G, Rajapaksha S, Bandara C (2013) Morphological, antimicrobial, durability, and physical properties of untreated and treated textiles using silver-nanoparticles. Colloids Surf. A: Physicochem. Eng. Aspects 436:975–989
26. Razvan S, Floarea P, Donciu C, Eftalea C, Carmen G, Maria B, Marcela R, Petronela D, Corina MI (2013) Contributions to the production of new types of knitted textile products with functional bioactive and conductive properties. In: Textile science and economy—5th international scientific-professional conference. Nov 05–06, 2013, Zrenjanin, Serbia
27. Rossin KJ (2010) Biomimicry: nature's design process versus the designer's process. In: Brebbia CA, Carpi A (ed) Design & nature V. WIT Press
28. Seydibeyoğlu MO, Oksman K (2008) Novel nanocomposites based on polyurethane and micro fibrillated cellulose. Compos Sci Technol 68(3–4):908–914
29. Srinivasarao M (1999) Nano-optics in the biological world: beetles, butterflies, birds, and moths. Chem Rev 99:1935–1961
30. Tejin (2006) Tejin Group CSR Report. http://www.teijin.com/rd
31. Teule F, Furin WA, Cooper AR, Duncan JR, Lewis RV (2007) Modifications of spider silk sequences in an attempt to control the mechanical properties of the synthetic fibers. Philos Trans R Soc 42:8974–8985
32. Tokatli N, Wrigley N, Kizilgun O (2008) Shifting global supply networks and fast fashion: made in Turkey for marks & spencer. Global Netw: A J Trans Affairs 8(3):261–280
33. Troward T (2007) The wisdom of thomas troward, vol I. Wilder Publications
34. Turley DB, Copeland JE, Horne M, Blackburn RS, Stott E, Laybourn SR, Harwood J (2009) The role and business case for existing and emerging fibres in sustainable clothing: final report to the Department for Environment, Food and Rural Affairs (Defra), London, UK
35. Vollarth F (2006) Spider silk: thousands of nano-filaments and dollops of sticky glue. Curr Biol 16(21):925–927
36. Vollrath F, Knight DP (2001) Liquid crystalline spinning of spider silk. Nature 410 (6828):541–548
37. Wagner P, Furstner R, Barthlott W, Neinhuis C (2003) Quantitative assessment to the structural basis of water repellency in natural and technical surfaces. J Exp Bot 54:1295–1303

Author Biographies

Geetha Dissanayake is currently a senior lecturer at the Department of Textile & Clothing Technology, University of Moratuwa, Sri Lanka. She has obtained her Ph.D. in Sustainable Consumption from the University of Manchester, UK, and the B.Sc Engineering in Textile & Clothing Technology from the University of Moratuwa, Sri Lanka. Her research interests includes sustainable textile and fashion, textile waste management, and remanufacturing.

Srimala Perera is working as a senior lecturer at the Division of Polymer, Textile and Chemical Engineering Technology, University of Moratuwa, Sri Lanka. She earned her Ph.D. in Nanotechnology from University of Peradaniya and the B.Sc Engineering in Textile & Clothing Technology from the University of Moratuwa, Sri Lanka. Her research focus is on nanotechnology in textiles and biomimicking.

Consumer Perceptions of Fibers with Respect to Luxury and Sustainability: An Exploratory Study

Cathy A. Rusinko and Marie-Eve Faust

Abstract This chapter contributes to the literature by examining an under-researched area: consumer perceptions of fibres with respect to luxury and sustainability. Theory-grounded questions are proposed, and an exploratory survey is designed and administered to a sample of young consumers. Results suggest that young consumers tend to perceive most fibres as *either* luxurious *or* sustainable and —with the exception of two fibers—*not both* luxurious *and* sustainable. In addition, sustainability was ranked last in importance with respect to characteristics deemed as important by consumers when considering fibres in clothing and apparel decisions. Findings have implications for how fibres—and the clothing and apparel made from them—can be designated and marketed.

1 Introduction

Over the past several years a growing body of literature has been emerging on consumer perceptions of luxury and sustainability with respect to fashion. In these studies, fashion has been defined in terms of clothing and apparel [1, 6, 22]. While sustainability and consumer perceptions of sustainability are being explored in the fashion luxury sector [15], little research exists on consumer perceptions of luxury and sustainability with respect to fibres, which are also defined as fashion goods [19]. Therefore, the purpose of this chapter is to contribute to the literature in this area and, in particular, to examine consumer perceptions of luxury and sus-

C.A. Rusinko (✉)
School of Business Administration, Kanbar College of Design Engineering & Commerce,
Philadelphia University, School House Lane and Henry Avenue,
Philadelphia, PA 19144, USA
e-mail: RusinkoC@PhilaU.edu

M.-E. Faust
École supérieure de Mode, École des sciences de la gestion,
Université du Québec à Montréal, Local DM-1315, Montreal, Canada
e-mail: faust.marie-eve@uqam.ca

© Springer Science+Business Media Singapore 2016
S.S. Muthu and M.A. Gardetti (eds.), *Sustainable Fibres for Fashion Industry*,
Environmental Footprints and Eco-design of Products and Processes,
DOI 10.1007/978-981-10-0566-4_2

tainability with respect to natural animal hair fibres. The chapter also contributes to the literature by examining whether consumers perceive the same fibres to be both luxurious and sustainable. Results of the study can contribute to the state of the art on how fibers—and the clothing and apparel that are made from them—are perceived by consumers, and how fibers, clothing and apparel can be marketed. Study results also have implications for usage of the terms "luxury" and "sustainability" with respect to fibres, clothing, and apparel.

The chapter begins by drawing upon the more recent literature to examine definitions of luxury and sustainability, and to summarize current findings on consumer perceptions of luxury and sustainable fashion. Then, questions—that are grounded in the literature on clothing and apparel—are proposed about consumer perceptions of natural animal hair fibres with respect to luxury and sustainability. Since our topic is under-researched, with little existing data or bases for hypotheses, we choose to propose exploratory questions rather than propose hypotheses. An exploratory study is designed and administered to address our questions and learn more about consumer perceptions of luxury and sustainability with respect to animal hair fibres. The data are analyzed and discussed and the limits of this exploratory study are considered. Finally, recommendations for future research are discussed.

2 Recent Literature

Both "luxury" and "luxury fashion" have been defined in several different ways [8, 17, 19]. One of the earliest and most widely cited luxury scales was developed by Kapferer [14]; he identified four different types of luxury consumers—those who look for: (1) beauty and excellence; (2) creativity and innovation; (3) rareness; and (4) brand reputation and timelessness [17]. DeBarrier et al. [3] combined the three most quoted luxury scales: Kapferer [14], Dubois et al. [4] and Vigneron and Johnson [26]. From these three scales, they derived and proposed a hybrid scale of eight dimensions of luxury: (1) elitism; (2) distinction and status; (3) rarity; (4) reputation; (5) creativity; (6) power of the brand; (7) hedonism; and (8) refinement. According to Dubois and Laurent [19], when luxury is linked to fashion goods, it often falls into categories—such as fur and leather goods, perfume, jewelry, apparel, and raw materials such as fine fibres, gold and diamonds. For example, Franck's [9] list of luxury animal hair fibres includes mohair, cashmere, pashmina (fine type of cashmere wool), camel hair, alpaca, llama, vicuna, angora and qiviut (the underwool of the Arctic muskox).

Both "sustainability" and "sustainable fashion" have been defined in several different ways, and researchers have cited the large number of definitions as problematic [1, 6]. One of the oldest and most cited definitions of sustainability is that offered by the Brundtland Commission of the United Nations in 1987. Their definition of sustainability or sustainable development is very broad, and can be paraphrased as that which meets the needs of the present generation without compromising the ability of future generations to meet their needs [27].

Another widely cited, more specific and more easily applicable perspective on sustainability is the triple bottom line (TBL). This study uses the concept of TBL to define sustainable fibres, and this definition will be presented and further explained later in the chapter. TBL was developed by Elkington [5] as a decision-making framework for sustainability in organizations, but has been extended to include decisions about sustainability in products, services and other entities. TBL holds that impacts with respect to three dimensions—economic/financial, social and environmental—must be considered equally in the decision-making process. This perspective differs from more traditional organizational decision making which typically considers mainly economic/financial impacts.

According to TBL, the impacts of any decision must be considered in terms of: (1) economic or financial sustainability, or the impact on the long-run survival of the organization, product, service, or entity; (2) social sustainability, or the impact on all members of society, including all members of the relevant supply chain plus other stakeholders (e.g., people living near manufacturing and/or disposal centers; future generations); and (3) environmental sustainability, or the impact on supply of natural resources and waste. TBL is sometimes referred to as the three Ps—profits, people and planets. Hence, makers of sustainable fibres must consider the impact of their supply chain with respect to: (1) economic or financial sustainability, or the long-run survival of their organization; (2) social sustainability, or all stakeholders in the supply chain, plus members of society in general; and (3) environmental sustainability, or use, waste and pollution of resources, including soil, air, water and other natural resources. Both the Brundtland Commission's definition of sustainability and TBL have been used by researchers in fashion and luxury [16, 10].

"Sustainable fashion" is often defined in terms of what it is not; that is, it is not fast fashion [1, 6]. Finn [6] observes that luxury fashion may have characteristics in common with sustainable fashion, since luxury clothing is more often passed down from generation to generation and not discarded. In Finn's words, "...amongst the mountains of clothing that are being discarded, why are there not more Channels, Vuittons or Saint Laurents?" (p. 2). Annamma et al. [1] suggest that luxury "can be both green and gold" (p. 289). They continue, "Luxury brands can become the leaders in sustainability because of their emphasis on artisanal quality" (p. 291). According to Van Nes and Cramer [25], consumers listed their primary requirements from eco-fashion to be durability, quality and style. According to Annamma et al. [1], durability, quality and style are also qualities of luxury brands. Hennigs et al. [13] reported that luxury brands are increasingly adopting sustainability as part of "the luxury essence" (p. 27). The majority of Kapferer and Michaut's [15] respondents reported that "luxury brands should be exemplary in terms of sustainability ... and, given the price of luxury, it should be shocking to learn that these brands are not compliant" (p. 12). However, an empirical study by Kapferer and Michaut-Denizeau [16] suggested that luxury buyers are somewhat ambivalent, in that they consider luxury and sustainability to be contradictory.

Some consumers—especially young, fast-fashion consumers—often perceive sustainable fashion to be unstylish and undesirable [1]. In addition, Annamma et al. [1] found that a large number of young people support sustainability through purchasing sustainable products and services in other aspects of their lives, but separate sustainability from their fashion purchases. Furthermore, older consumers, such as female professionals who purchase clothing for themselves and their children, supported the idea of sustainability, but did not purchase sustainable clothing due to price and convenience issues (Ritch and Schroder [22]).

Luxury fibres have been defined and listed in several sources, with some of the most recent compilations by Karthik et al. [18] and Hassan [12]. Karthik et al. [18] include natural, luxury fibres from animals, insects and plants, and Hassan [12] includes natural fibres from animals and insects. Since our study is exploratory, we focused on nine natural, luxury fibres from animals, in the interest of keeping our focus limited. We drew from Franck's [9] list of natural animal hair fibres, which is generally consistent with the lists by Hassan [12] and Karthik et al. [18]. However, future studies should include luxury insect and plant fibres, as well as additional animal fibres—such as cervelt (rare natural down fibre from red deer) and guanaco —and this will be discussed later in the chapter.

3 Research Questions

The differing perspectives reflected above point to yet unresolved questions about consumer perceptions of luxury fashion, sustainable fashion and the relationship between luxury fashion and sustainable fashion. To shed further light on these issues, we propose four exploratory questions. Furthermore, we examine an under-researched topic within the fashion luxury products category, which is luxury fibres [7], since they are included in the category of fashion luxury products [19]. Since there is little research on consumer perceptions of luxury and sustainable fibres, our questions are grounded in recent research on this topic with respect to other luxury and sustainable fashion products. Our questions include:

Q1. Do consumers perceive fibres that are designated by experts as "luxury fibres" to be luxurious?

Q2. Why do consumers perceive a particular fibre as luxurious—or not luxurious?

The first question will help to determine whether consumer perceptions of luxury fibres are consistent with expert definitions of luxury fibers. Since researchers have cited definitional problems with the term "luxury", due to its subjective nature (Kapferer and Michaut-Denizeau [16]), it is important to know whether consumers define luxury fibres in the same way as the experts and researchers who develop the scales and lists. For example, Kapferer and Michaut-Denizeau [16] discussed the history of luxury, which was allied with rare, high-quality products that were made by hand, and with respect for tradition. They contrast this with the current era of mass luxury, whereby luxury may better resemble consumer or fashion goods, which will quickly become obsolete and discarded. Hence, the answer to the first question can provide insight about consumer perceptions of luxury fibers, or the lack thereof, and may have implications for how luxury and luxury fibers are designated and marketed in the future.

The second question will help to determine whether consumer perceptions of luxury fibres are consistent with consumer perceptions of other types of luxury goods, as specified by the three most often quoted luxury scales: Kapferer [14], Dubois et al. [4], and Vigneron and Johnson [26]. For example, Kapferer [17] identified four different types of luxury consumers—those who look for (1) beauty and excellence; (2) creativity and innovation; (3) rareness; and (4) brand reputation and timelessness. DeBarrier et al. (2012) combined the three most quoted luxury scales (listed above) and derived a hybrid scale of eight dimensions of luxury: (1) elitism; (2) distinction and status; (3) rarity; (4) reputation; (5) creativity; (6) power of the brand; (7) hedonism; and (8) refinement. Since these luxury scales address luxury goods in general, and not luxury fibres per se, it is important to investigate consumer perceptions of luxury fibres with respect to these scales. The answer to this question may have implications for how luxury fibres are designated and marketed in the future.

Based on the literature discussed above, we can ask additional questions with respect to consumer perceptions of luxury fibres:

Q3: Do consumers perceive luxury fibres to be sustainable?

Q4: What variables influence consumer choices of fibres when shopping for fashion?

With respect to the third question, some researchers [2, 6] have suggested that luxury fashion and sustainable fashion have some characteristics in common. Annamma et al. [1] suggest that luxury "can be both green and gold" (p. 289). Likewise, Hennigs et al. [13] reported that luxury brands are increasingly adopting sustainability as part of the experience or value of their product. However, Kapferer and Michaut-Denizeau [16] reported that luxury customers are ambivalent, since they find luxury and sustainability to be contradictory. The third question will help to determine whether consumers perceive a relationship between luxury and sustainability with respect to fibers. The answer to this question may have implications for how luxury and sustainable fibers are designated and marketed in the future.

With respect to the fourth question, some researchers (Ritch and Schroder [22]) have suggested that variables such as sustainability are least important—and characteristics such as price and quality are most important—when consumers shop for clothing and apparel. Furthermore, older consumers, such as female professionals who purchase clothing for themselves and their children, supported the idea of sustainability, but did not purchase sustainable clothing due to price and convenience issues (Ritch and Schroder [22]). Annamma et al. [1] found that a large number of young people support sustainability by purchasing sustainable products and services in other aspects of their lives, but separate sustainability from their fashion purchases. The fourth question will help to determine which variables are most and least important with respect to fibres in clothing and apparel when consumers are making purchasing decisions. The answer to this question may have implications for how fibers are designated and marketed in the future.

4 Methods

To address the four questions above, we designed an exploratory study and administered it through hard-copy (paper) survey instruments.

To address Q1, which set out to examine consumer perceptions of luxuriousness of natural animal hair fibres, we used the luxury animal hair fibres in Franck's [9] list of luxury natural animal hair fibres: alpaca, angora, camel hair, cashmere, llama, mohair, pashmina, qiviut, and vicuna. Frank's list was generally consistent with more recent compilations of natural fibres, such as Hassan [12] and Karthik et al. [18]. We chose Franck's list since it enumerated animal hair fibres only, unlike the more recent lists that included insect and plant fibres [12, 18]. We focused on luxury natural animal hair fibres only, as opposed to including luxury insect or vegetable fibres, in order to keep this exploratory study manageable and to a manageable length to maximize responses. Future research should include luxury insect and vegetable fibres; this will be discussed later in this chapter.

In addition to Frank's (2001) list of luxury natural animal hair fibres, we added merino wool to our list, giving a total of 10 types of natural animal fibres. We added merino wool due to recent efforts by industry groups and designers to recognize and use merino wool as a luxury fibre [20]. For example, The No Finer Feeling ad campaign informs consumers about the natural benefits and attributes of merino wool. Fashion leaders, including Vivienne Westwood, have become ambassadors for the initiative and for merino wool as a luxury fibre. Hence, this study will measure consumer perceptions of merino wool with respect to luxury.

For Q1, we listed the 10 types of luxury natural animal fibres with a Likert-type scale of choices and asked respondents to check the one response that described how luxurious they considered each type of fiber to be: not luxurious, a little luxurious, luxurious, very luxurious or unfamiliar with this fibre. Q1 and results are illustrated in Table 1 and discussed in the next section, "Results".

For Q2, we set out to examine consumer perceptions of why fibers are luxurious. We drew upon Kapferer's luxury scale, which describes four types of luxury customers, or those who look for: (1) beauty and excellence; (2) creativity and innovation; (3) rareness; and (4) brand reputation and timelessness. We also drew upon the hybrid, eight-variable luxury scale developed by DeBarnier et al. [3] which is based upon the three most cited luxury scales: Kapferer [14], Dubois et al. [4] and Vigneron and Johnson [26]. Drawing upon these luxury scales—and in order to make the survey easily understandable by respondents—we included six different categories as possible explanations for why consumers perceive fibres to be luxurious. Respondents were instructed to check as many boxes as applied. It is important to note that our six categories map onto both Kapferer's luxury scale and

DeBarnier's hybrid luxury scale, as illustrated in Appendix. Our six categories to explain consumer perceptions of why fibres are luxurious are: (1) Costly (economically expensive); (2) Rare/Unique (scarce, limited, uncommon); (3) Good Feel (on the skin/body); (4) Status Symbol (respected, admired, indicative of high social standing); (5) Beautiful (pleasing to the senses or mind); and (6) Always in Style (timeless).

For Q2, we listed the 10 natural animal fibres from Q1 and asked respondents to check all categories that applied to explain why they considered each fibre to be luxurious. In addition to the six categories listed above, we also added a category of N/A, if respondents were unfamiliar with the fibre. Q2 and results are illustrated in Table 2 and discussed in the next section, "Results".

To address Q3, which sets out to examine whether consumers perceived luxury fibres as sustainable, we used the list of 10 natural animal fibres from Q1 and Q2. We also used a Likert-type scale of choices and asked respondents to check the one response that described how sustainable they considered each type of fiber to be: not sustainable, a little sustainable, sustainable, very sustainable or unfamiliar with this fiber. We provided the following definition of sustainable fibres: "Sustainable fibers can be defined as fibres that are produced with minimal negative impact on the environment and society". For example, they use resources efficiently during production and processing (minimal waste), avoid use of harmful chemicals during production and processing and treat workers and animals ethically (no sweatshops or animal cruelty). This definition includes applications of economic, environmental and social sustainability and, hence, captures the TBL perspective on sustainability that was explained earlier in this chapter. The first part of the definition (efficient use of resources and minimal waste) most strongly addresses economic and environmental sustainability. The next part of the definition (avoiding use of harmful chemicals) most strongly addresses environmental sustainability. The last part of the definition (treating workers and animals ethically) most strongly addresses social sustainability. The definition is purposely brief in the interest of conciseness and maximizing the number of survey responses.

In addressing Q3, we are testing whether consumers perceive luxury fibres as both luxurious and sustainable; this issue is currently being explored in the literature and, thus far, findings are not conclusive. For example, Finn [6] observed that luxury fashion may have characteristics in common with sustainable fashion, since luxury clothing is more often passed down from generation to generation and not discarded. Annamma et al. [1] suggested that luxury "can be both green and gold" (p. 289). According to Van Nes and Cramer [25], consumers listed their primary requirements from eco-fashion to be durability, quality and style. According to Annamma et al. [1], durability, quality and style are also qualities of luxury brands. Hennigs et al. [13] reported that luxury brands are increasingly adopting sustainability as part of "the luxury essence" (p. 27). The majority of Kapferer and Michaut's [15] respondents reported that "luxury brands should be exemplary in terms of sustainability ... and given the price of luxury, it should be shocking to learn that these brands are not compliant" (p. 12). However, an empirical study by Kapferer and Michaut-Denizeau [16] suggested that luxury buyers are somewhat ambivalent, in that they consider luxury and sustainability to be contradictory. Q3 and results are illustrated in Table 3 and discussed in the next section, "Results".

To address Q4, which sets out to examine consumer ranking of importance of variables when choosing fibres for clothing and accessories, we used a list of nine variables, and asked respondents to number the variables from 1 to 9, to reflect their importance in considering fibres, when shopping for clothing and accessories (1 was most important and 9 was least important). Four of the variables for Q4 were drawn from our luxury scale from Q2 (attractiveness, feel/comfort, rareness/uniqueness, status). We also used the variables of price, quality, durability and whether the product was in style at the time (of choice) since these variables were perceived as important in earlier studies to explain clothing choices by consumers [22]. We also added the variable of sustainability, so that we could measure the importance of sustainability when consumers choose fibres for clothing and accessories. We used the same definition of sustainability/sustainable fibres as was used in Q3. Therefore, the variables for Q4 were price, quality, durability, attractiveness, whether it is in style at the time, feel/comfort, rareness/uniqueness, status and sustainability (meaning sustainable fibre). Q4 and results are illustrated in Table 4 and discussed in the next section, "Results".

The sample of respondents consisted of 45 undergraduate students in fashion merchandising and fashion design classes, who ranged in age from 17 to 23 years old. While these students may be more knowledgeable about fashion, luxury and sustainability than students from other majors, they represent future fashion professionals and/or the most engaged and knowledgeable fashion consumers. They also represent the Millennial Generation, which is the up-and-coming generation whose economic spending power will be growing over the next several decades. For these reasons, they are a good sample for our study, which is an initial, exploratory study. We had 42 complete responses to Q1, Q2 and Q4 of the anonymous survey—a response rate of 93 %. We had 36 complete responses to Q2 of the anonymous survey—a response rate of 86 %.

5   Results and Discussion

Results of the survey are reported in the tables below and are further explained below. Results in the form of answers to the four questions proposed in this study are also provided, as is discussion.

Results of Table 1
For Table 1 the numbers in each column represent the total number of respondents ($n = 42$) and the numbers in brackets in column V represent the total percentage of respondents ($n = 42$). As indicated in Table 1, consumers perceive some luxury fibres as more luxurious than others. Cashmere is clearly perceived as the most luxurious fibre, with 90 % of respondents rating it as luxurious or very luxurious. All respondents were familiar with cashmere, as well as llama and merino wool.

Table 1 (Q1): Consumer perceptions of luxuriousness of natural animal hair fibres (check one box)

Type of fiber	I. Not luxurious	II. A little luxurious	III. Luxurious	IV. Very luxurious	V. III. + IV.	VI. Unfamiliar with this fibre
Alpaca	1	15	20	4	24 [57]	2
Angora	2	8	16	8	24 [57]	8
Camel hair	5	16	9	8	17 [40]	2
Cashmere	0	4	17	21	38 [90]	0
Llama	8	14	12	4	16 [38]	0
Mohair	2	13	13	7	20 [48]	5
Pashmina	1	10	15	6	21 [50]	10
Qiviut	1	2	4	9	13 [31]	26
Vicuna	2	3	7	14	21 [50]	15
Merino wool	9	17	9	7	16 [38]	0

Note Numbers in each column represent the total number of respondents providing that answer ($n = 42$), and the numbers in brackets in column V represent the total percentage of respondents indicated by the non-bracketed number in column V ($n = 42$)

After cashmere, alpaca and angora were rated as luxurious or very luxurious by 57 % of respondents. Pashmina and vicuna were rated as luxurious or very luxurious by 50 % of respondents. In the case of pashmina and vicuna, 24 % and 36 % of respondents, respectively, were unfamiliar with the fibre. Qiviut was the least familiar fibre, with 62 % of respondents reporting that they were unfamiliar with it.

In contrast, all respondents were familiar with merino wool; however, only 38 % rated it as luxurious or very luxurious and the highest percentage of respondents (21 %) rated it as not luxurious. This may be because merino wool has not enjoyed traditional status as a luxury fibre, and efforts by the industry to label it as such are relatively new [20].

Results of Table 2
In Table 2 the number in each column represents the total number of respondents ($n = 42$). Table 2 illustrates why consumers in our sample perceived fibres to be luxurious. Cashmere, which was rated the most luxurious fibre in Table 1, received clearly the highest number of respondent ratings with respect to why it was perceived as luxurious. Consumers perceived cashmere to be luxurious because of its costliness, good feel, status, beauty, and consistent stylishness. Hence, in this sample, consumer perceptions of why cashmere is luxurious reinforced earlier rationales by Kapferer [17] and others for why consumers perceive goods as luxurious.

On the other hand, cashmere received much lower respondent ratings for rareness/uniqueness; in fact, cashmere received the second lowest respondent rating for rareness/uniqueness after merino wool. As reported in Table 1, 38 % of respondents rated merino wool as luxurious or very luxurious, while 90 % of respondents rated cashmere as luxurious or very luxurious. Therefore, this group of respondents perceives cashmere as an extremely luxurious fibre—for reasons of its

Table 2 (Q2): Consumer perceptions of why fibres are luxurious (check as many boxes as apply)

Type of fibre	Costly	Rare/Unique	Good feel	Status symbol	Beautiful	Always in style	N/A
Alpaca	16	10	20	6	16	5	5
Angora	16	12	18	6	15	3	11
Camel hair	12	20	7	6	12	3	5
Cashmere	32	6	32	28	30	30	1
Llama	5	12	8	4	11	1	6
Mohair	14	12	16	5	12	2	7
Pashmina	18	10	20	9	19	8	10
Qiviut	11	14	7	9	8	2	26
Vicuna	14	15	8	9	6	2	15
Merino wool	11	2	20	7	21	24	3

Note Numbers in each column represent the total number of respondents providing that answer (*n* = 42)

costliness, good feel, status, beauty and timelessness with respect to style—and their overall perception of the luxuriousness of cashmere is not negatively affected by their corresponding perception that cashmere is not a rare or unique fibre.

As indicated in Table 2, consumers in this sample tended to respond strongly with respect to cost, good feel and beauty when explaining why fibres were perceived as luxurious. These results for fibres echo results for other types of luxury goods [17].

Results of Table 3

In Table 3 the numbers in each column represent the total number of respondents (*n* = 36). The largest number of respondents (26 respondents or 72 %) perceived merino wool to be a sustainable or very sustainable fibre. Alpaca was perceived a

Table 3 (Q3): Consumer perception of luxury fibres as sustainable (check one box)

	I. Not sustainable	II. A little sustainable	III. Sustainable	IV. Very sustainable	V. III. + IV.	VI. Unfamiliar with fibre
Alpaca	5	5	17	8	25 [69]	1
Angora	8	7	9	8	17 [47]	3
Camel hair	9	8	12	6	18 [50]	1
Cashmere	7	8	13	5	18 [50]	0
Llama	7	2	17	7	24 [67]	0
Mohair	6	5	13	7	20 [56]	3
Pashmina	8	7	7	4	11 [31]	10
Qiviut	6	4	5	3	8 [22]	18
Vicuna	8	6	6	3	9 [25]	13
Wool (merino)	5	4	19	7	26 [72]	1

Note Numbers in each column represent the total number of respondents providing that answer (*n* = 36), and the numbers in brackets in column V represent the total percentage of respondents indicated by the non-bracketed number in column V (*n* = 36)

Table 4 (Q4): Consumer ranking of importance of nine variables when considering which fibres to choose while shopping for clothing and accessories (1 is most important and 9 is least important)

Variables/Importance	1	2	3	4	5	6	7	8	9
Price	11	7	7	7	4	4	1	1	0
Quality	9	5	7	10	6	5	0	0	0
Durability	0	4	6	5	10	10	4	3	0
Attractiveness	20	7	6	4	2	2	0	1	0
Whether it is in style at the time	2	2	5	2	7	6	11	6	1
Feel/Comfort	5	12	6	5	5	3	3	2	1
Rareness/Uniqueness	1	4	5	2	4	6	5	13	2
Status	0	1	0	3	0	3	8	9	18
Sustainability	0	0	2	3	4	3	7	5	18

Note Numbers in each column represent the total number of respondents providing that answer ($n = 42$)

sustainable or very sustainable fibre by 69 % of respondents, with the perception of llama as a sustainable or very sustainable fibre by 67 % of respondents. Other fibers that were perceived as sustainable or very sustainable by at least 50 % of respondents were mohair (56 %), camel hair (50 %) and cashmere (50 %).

Results of Table 4
In Table 4 the number in each column represents the total number of respondents ($n = 42$). The largest number of respondents (20 respondents or 48 %) rated attractiveness as the most important variable (#1) when considering which fibers to choose while shopping for clothing and accessories. The variable with the second largest number of #1 (most important) ratings was price (11 respondents, or 26 percent), and quality received the third largest number of number one ratings (9 respondents or 21 %). The fourth largest number of #1 ratings was feel/comfort (5 respondents or 12 %). These results for fibres echo results for other types of fashion [22] with respect to consumer perceptions of important variables when shopping.

Correspondingly, the largest number of respondents (18 respondents or 43 %) rated sustainability and status as the least important variable (#9) when considering which fibers to choose while shopping for clothing and accessories. The low ranking of sustainability with respect to its importance in fibre choices is consistent with earlier research on consumer perceptions of the importance of sustainability in clothing choices (Ritch and Schroder [22]; [1]).

5.1   Answers to the Study Questions

As stated above, this chapter proposed four exploratory questions about consumer perceptions of luxury and sustainability with respect to fibres. While the focus on fibres is an under-researched area the questions were grounded in earlier research on luxury and sustainable fashion. Measures were designed and a survey instrument was developed and administered to a sample of young fashion consumers. Based on survey responses, each question (and response) will be addressed and discussed here, and discussed further in the following section, "Discussion":

Q1: Do consumers perceive fibres that are designated by experts as "luxury fibers" to be luxurious?

As discussed in detail in the "Results" section, some fibres were perceived as much more luxurious than others. Cashmere was clearly perceived as the most luxurious by these respondents, with alpaca and angora ranked as distant seconds. Respondents were unfamiliar with several of the fibres. Therefore, our respondents do not tend to perceive luxury fibres similarly to experts in the field [9].

Q2: Why do consumers perceive a particular fibre as luxurious—or not luxurious?

As discussed in detail in the "Results" section, consumers tended to respond strongly with respect to cost, good feel and beauty when explaining why fibres were perceived as luxurious. These results for fibres echo results for other types of luxury goods [17].

Q3: Do consumers perceive luxury fibres to be sustainable?

As discussed in detail in the "Results" section the luxury fibres in this study were not generally perceived as both luxurious and sustainable by high percentages of respondents. Alpaca was the only fibre that was perceived as both luxurious or very luxurious and sustainable or very sustainable by more than 50 % of respondents.

Q4: What variables influence consumer choices about fibres when shopping for fashion?

As discussed in detail in the "Results" section, respondents rated attractiveness, price and quality as first, second and third in importance, respectively. The largest number of respondents rated sustainability as the least important variable (#9) as they consider which fibres to choose when shopping for clothing. These results for fibres echo the results for other types of fashion (Ritch and Schroder [22]) with respect to consumer perceptions of important variables when shopping.

6 Discussion

Since this was a small, exploratory study on an under-researched topic—perception of luxury and sustainability with respect to fibres—our ability to generalize our findings is limited. However, despite the limits of this study, it asked some new questions and yielded some interesting results that may inform future research and practice.

With respect to luxury, our results tend to reinforce some of the earlier research on consumer perceptions of other types of luxury fashion and products. For example, in our study, consumer explanations for why they perceived fibres as luxurious were most often costliness, beauty and good feel (on the body). These findings are consistent with earlier research and research frameworks for luxury goods [17].

However, when presented with a list of luxurious animal hair fibres, as articulated by Franck [9], respondents do not perceive all fibres on the list as luxurious. Only half the fibres on the list were perceived as luxurious or very luxurious by at least 50 % of respondents. This result may be partially due to lack of familiarity with some of the lesser known fibres (e.g., qiviut, llama). However, fewer than 50 % of respondents perceived some of the better known fibres (e.g., camel hair, mohair) as luxurious. These results—especially if replicated in a larger and more diverse survey sample—may have implications for how luxury fibres are defined and/or marketed in the future.

With respect to two of the best known fibres (cashmere and merino wool) in this study, there seems to be an inverse relationship between perceptions of luxury and perceptions of sustainability. That is, cashmere was perceived as luxurious or very luxurious by the highest percentage of respondents (90 %), but was perceived as sustainable or very sustainable by only 50 % of respondents. Merino wool was perceived as luxurious or very luxurious by 38 % of respondents, but perceived as sustainable or very sustainable by 72 % of respondents (the highest percentage ranking as a sustainable fibre among all fibres in the survey). Hence, none of the fibres in this study was perceived as both luxurious and sustainable by high percentages of respondents. This finding may be consistent with earlier findings (Annamma 2012) that younger consumers tend to perceive sustainable fashion as less desirable (and, by extension, may not perceive the same fashion or fibre as both sustainable and luxurious). This finding may also be consistent with the findings of Kapferer and Michaut-Denizeau [16]; they reported that luxury buyers consider luxury and sustainability to be somewhat contradictory. This relationship should be retested in a larger and more diverse sample of subjects.

In this study, it is also interesting that cashmere's high ranking as a luxurious fibre coexists with its relatively low ranking as a rare or unique fibre, especially since experts such as Kapferer [17] define one dimension of luxury as rareness or uniqueness. Perhaps this finding may be explained by the relatively plentiful supply of cashmere clothing and accessories over the past several years; however, this relationship should be retested in a larger and more diverse sample of subjects. Correspondingly, future studies should also examine the relationship between consumer perceptions of high volume (or plentiful supply) and luxury.

The only fibre that was perceived as both luxurious or very luxurious, and sustainable or very sustainable, by more than 50 % of respondents was alpaca. These results—especially if replicated in a larger and more diverse survey sample—may have implications for how fibres are designated and marketed and, in particular, how alpaca, cashmere and merino wool are designated and marketed in the future. For example, alpaca may be able to be designated and marketed as both luxurious and sustainable, especially to young consumers who made up this sample. This type of marketing may attract young, fast-fashion consumers, who are supportive of sustainability across their lifestyles, but do not buy sustainable or eco-fashion because they perceive it as unattractive or unfashionable [1]. Likewise, this type of marketing may also attract consumers who oppose fast fashion but support sustainable fashion choices. Of course, if marketers decide to market alpaca as a fibre that is both luxurious and sustainable, they will have to address and integrate sustainable practices along the whole value chain—or risk accusations of greenwashing, and subsequent negative consequences.

Given current consumer perceptions of merino wool as much more sustainable than luxurious—and recent efforts by the industry to market it as luxurious—the industry might consider strategies to change perceptions of merino wool so that consumers see it as both a luxury fibre and a natural/sustainable fibre. If these strategies are successful, merino wool may be able to be marketed to both luxury consumers and consumers who value sustainable fashion. Alternatively—and/or if strategies to market merino wool as a luxury fiber fall short of expectations—the industry may want to leverage current perceptions of merino wool as a sustainable fibre.

Based on the results of this exploratory study, consumers tend to choose fibres that are attractive, economical and high quality. If these results can be replicated with larger, more diverse samples, the industry may want to consider an educational campaign that focuses on the attractiveness, economy and quality of sustainable fibres. In addition, the industry may consider educational campaigns to better familiarize consumers with lesser known natural fibres such as qiviut and vicuna.

7   Conclusions and Recommendations for Future Research

The results of this study suggest that young consumers do not tend to define luxury animal hair fibres in the same way as the experts who construct the lists of luxury animal hair fibres. They also suggest that young consumers do not tend to perceive luxuriousness of fibres in the same way as the experts who construct the fashion luxury scales. Young consumers identify more strongly with the qualities of cost, good feel and beauty, in general. However, these same young consumers identify less strongly with the qualities of rareness/uniqueness, status symbol and always in style. While these findings may be at least partly a function of the relatively young age (17- to 23-year-olds) and the relatively small sample of respondents in this study, future studies should continue to investigate these issues and the corresponding implications for defining, labelling and marketing luxury animal hair fibres to the up-and-coming generation of millennial consumers represented in this study, in particular. According to the Pew Research Foundation, the Millennial Generation currently ranges in age from about 15 to 34 years old (those born from about 1980 to about 2000), so the economic purchasing power of this group will continue to grow over the next several decades.

These study results also suggest that young consumers tend to categorize luxury animal fibres as either luxurious or sustainable—and not both luxurious and sustainable. These results may confirm findings by other researchers, such as Annamma et al. [1], who reported that young, fast-fashion consumers perceive sustainable fashion to be undesirable and unstylish, and Kapferer and

Michaut-Denizeau [16], who reported that luxury buyers tend to view luxury and sustainability as contradictory. To follow up these results, future studies should continue to examine which fibres are considered luxurious, which are considered sustainable, and which are considered both—and why. In addition to the luxury animal fibres included in this study, other luxury animal fibres such as cervelt and guanaco should be included in future research. In addition to luxury animal fibres, luxury fibres from insects (silks) and plants (e.g., lotus, milkweed, pine, pineapple leaf, soy, and others) should also be included in future research.

Future studies should also employ diverse research methods. Richer methods such as interviews can help to uncover why, for example, cashmere receives high respondent ratings with respect to luxuriousness, despite the fact that consumers do not perceive it to be unique or rare. Follow-up surveys and richer research methods can also help to explain whether there is an inverse relationship between perceptions of luxury and perceptions of sustainability with respect to fibres—and better specify the relationship, if one exists.

With respect to young consumer rankings of variables when they are considering which fibers to choose for clothing and accessory purchases, respondents ranked price, attractiveness and quality as first, second and third, respectively. Sustainability was ranked last (#9). These results may confirm similar findings by earlier researchers with respect to other fashion goods (Ritch and Schroder [22]). While the results in this study may be partially due to the relatively young age and budget consciousness of respondents, they may also confirm earlier research by researchers such as Annamma et al. [1] that suggests that young people tend to support sustainability through purchases in other aspects of their lives, but separate it from their fashion purchases, perhaps due to the perception of sustainable fashion as unattractive and/or unstylish. It is vital that future studies more closely study and analyze these issues, particularly since they apply to the Millennial Generation, whose economic purchasing power will continue to grow in the future.

While this study focuses on consumer perceptions, future studies should also address sustainable practices and opportunities across entire fibre supply chains. Gardetti and Muthu's [11] examination and illustration of a business model canvas for sustainable apparel can serve as a template for similar application by fibre supply chains.

Equally importantly, the fashion industry needs to agree upon more standardized definitions for the terms "luxury" and "sustainability/sustainable". Research on sustainability from other industries [23, 24] has suggested that lack of a common industry definition of sustainability can confuse customers and, correspondingly, can dilute rewards (e.g., winning green customers) for industry members who pursue legitimate sustainable practices. In addition, Rusinko's findings suggest that lack of a common industry definition for sustainability can hamper efforts to expose industry members who are greenwashing. Similar arguments can be made for a more standardized definition of the term sustainability in the fashion industry. In addition, a more standardized definition of sustainability can serve to educate customers, and customers who better understand sustainable fashion and sustainable fibres may be less likely to perceive them as unattractive and/or unstylish.

The same can be said with respect to a more standardized definition of luxury in the fashion industry. That is, a more common definition (or set of definitions) for luxury in the fashion industry may not only reward members who are offering legitimate luxury goods, but may also act to expose those who are offering counterfeit goods [21]. A set of common definitions can be as broad and encompassing as is desirable by industry members.

In the future, surveys to address the questions proposed in this study should be administered to larger samples and to more diverse samples. For example, both focussed samples of luxury buyers and focussed samples of sustainable buyers should be surveyed, as well as broader, more general samples of fashion buyers. Since the sample surveyed for this exploratory study was young (17- to 23-year-olds), future samples should include older buyers, and/or a more general cross-section of buyers across all age groups. Since the sample surveyed for this exploratory study was relatively small, future studies should survey larger groups of buyers.

Appendix: Proposed Definitional Mapping of Rusinko and Faust's Luxury Scale (Adapted from Kapferer) onto a Hybrid Luxury Scale (DeBarnier et al.) Based on the Three Most Quoted Luxury Lcales

"Across the top" should not appear in print–it is directing the editor/typesetter about where "Six Luxury Variables Used by Rusinko and Faust (adapted from Kpaferer)" should be placed.

"Down the side" should not appear in print–it is directing the editor/typesetter about where "Eight Luxury Varialbles by DeBarnier et al." should be placed.

Luxury variable	Costly	Rare/Unique	Good feel	Status symbol	Beautiful	Always in style
Elitism	×	×				
Distinction and status	×			×		×
Rarity		×				
Reputation				×		
Creativity		×				
Power of the brand				×		×
Hedonism			×			
Refinement					×	×

Note Mapping is proposed to illustrate that the luxury scale variables used by Rusinko and Faust are grounded in existing theory and existing luxury scales

Down the side: Eight luxury variables by DeBarnier et al.

References

1. Annamma SJ Jr, Venkatesh A et al (2012) Fast fashion, sustainability, and the ethical appeal of luxury brands. Fashion Theory 16:273–296
2. Castelli CM, Brun A (2012) What is luxury in our society: investigation of the concept of luxury from a consumer market point of view. In: Paper presented at the international workshop on luxury retail, operations and supply chain management, Politecnico di Milano, Milan, Italy, 3–4 Dec 2012
3. DeBarnier V, Falcy S et al (2012) Do consumers perceive three levels of luxury? A compariason of accessible, intermediate and inaccessible luxury brands. J Brand Manage 19:623–636
4. Dubois B, Laurent G et al (2001) Consumer rapport to luxury: Analyzing complex and ambivalent attitudes. HEC, Jouy en Josas, France. Consumer research working paper no. 736
5. Elkington J (1998) Cannibals with forks: the triple bottom line of 21st century business. New Society Publishers, Gabriola Island, BC
6. Finn AL (2011) Luxury fashion: the role of innovation as a key contributing factor in the development of luxury fashion goods and sustainable fashion design. In: Satochi (ed) proceedings of the conference: fashion & luxury: between heritage and innovation, Institut Francais de la Mode (IFM), Paris, France, 18 Nov 2011
7. Faust ME (2013) The lux-story supply chain, told by retailers to build a competitive sustainable competitive advantage. Int J Retail Distrib Manage 41:973–985
8. Fionda AM, Moore CM (2009) The anatomy of the luxury fashion brand. Brand Manage 16:347–363
9. Franck RR (ed) (2001) Silk, mohair, cashmere and other luxury fibres. Woodhead, Cambridge
10. Gardetti MA (2015) Loewe: luxury and sustainable management. In: Gardetti MA, Muthu SS (eds) Handbook of sustainable luxury textiles and fashion, vol 2. Springer, Singapore, pp 1–16
11. Gardetti MA, Muthu SS (2015) Sustainable apparel? Is the innovation in the business model?— the case of the IOU project. Text Cloth Sustain. doi:10.1186/s40689-015-0003-0
12. Hasssan MM (2015) Sustainable processing of luxury textiles In: Gardetti MA, Muthu SS (eds) Handbook of sustainable luxury textiles and fashion, vol 1. Springer, Singapore, pp 101–120
13. Hennigs N, Wiedmann KP et al (2013) Sustainability as part of the luxury essence. J Corp Citizens 52:25–35
14. Kapferer JN (1998) Why are we seduced by luxury brands? J Brand Manag 16:290–301
15. Kapferer JN, Michaut A (2015) Luxury and sustainability: a common future: the match depends upon how consumers define luxury. Luxury Res J 1:3–17
16. Kapferer JN, Michaut-Denizeau A (2013) Is luxury compatible with sustainability? luxury consumers' viewpoint. J Brand Manage 21:1–22
17. Kapferer JN (2012) The luxury strategy: break the rules of marketing to build luxury brands, 2nd edn. Boca Raton, Kogan Page
18. Karthik T, Raithinamoorthy P et al (2015) Sustainable luxury natural fibers—production, properties, and prospects In: Gardetti MA, Muthu SS (eds) Handbook of sustainable luxury textiles and fashion, vol 1. Springer, Singapore, pp 59–100
19. Laurent G, Dubois B (1996) The functions of luxury: a situational approach to excursionism. Adv Consum Res 23:470–477
20. Lockwood L (2013) Narciso Rodriguez named Woolmark ambassador. Women's wear daily. 30 Jan 2013
21. Phau I, Teah M (2009) Devil wears (counterfeit) Prada: a study of antecedents and outcomes of attitudes towards counterfeits of luxury brands. J Consum Market 26:15–27
22. Ritch EL, Schroder MJ (2012) Accessing and affording sustainability: the experience of fashion consumption within young families. Int J Consum Stud 36:203–210
23. Rusinko CA (2010) Evolution of environmentally sustainable practices: the case of the U.S. carpet industry and CARE. Int J Sustain Econ 2:250–259

24. Rusinko CA (2007) Green manufacturing: an evaluation of environmentally sustainable manufacturing practices and their impact on competitive outcomes. IEEE Trans Eng Manage 54:445–454
25. Van Nes N, Cramer J (2005) Influencing product lifetime through product design. Bus Strategy Environ 14:286–299
26. Vigneron F, Johnson L (2004) Measuring perceptions of brand luxury. J Brand Manag 11:484–506
27. World Commission on Environment and Development (WoCD) (1987) Our common future. Oxford University, Oxford

Sustainable Natural Fibres from Animals, Plants and Agroindustrial Wastes—An Overview

Shahid-ul-Islam and Faqeer Mohammad

Abstract Natural fibres are produced by plants and animals and are the most important raw material for textile and non-textile applications. Natural fibres—particularly wool, cotton and silk—have been an integral part of human life and society since antiquity. With the discovery of synthetic fibres—such as nylons, acrylics and polyesters—the use of natural fibres decreased to a large extent. Recently (as of 2016), growing environmental awareness among the government and industrial firms has dramatically increased the worldwide market for all sustainable natural fibres. Lately, in addition to plant sources—cotton, jute, flax and ramie (a flowering plant in the nettle family)—and animal sources—wool and silk—agroindustrial wastes are exploited as viable, abundantly available, cheap and renewable alternative sources for natural fibres. The natural cellulose fibres recently extracted from corn stalks, corn husks, wheat straw, rice husk, sorghum stalk and leaves, banana leaves, pineapple leaves, sugarcane stalks, hop stems, soybean straw, etc. and protein fibres from industrial wastes have suitable composition, properties and structure for use in various textile and non-textile applications. This chapter first highlights the sources and important characteristics of wool, cotton, ramie and jute and finally discusses the production of cellulose and protein-based natural fibres from agricultural wastes.

1 Introduction

Synthetic fibres such as nylon, polyethylene terephthalate, polyamides and aramides offer several advantages—such as high strength, cheapness and low moisture absorption—for extensive use in different application fields notably in the textile industry [1, 2]. The textile industry utilizes a lot of chemicals in processes—such as sizing, scouring, bleaching, mercerizing (giving lustre), dyeing, printing and

Shahid-ul-Islam · F. Mohammad (✉)
Department of Chemistry, Jamia Millia Islamia, New Delhi 110025, India
e-mail: faqeermohammad@rediffmail.com

Shahid-ul-Islam
e-mail: shads.jmi@gmail.com

© Springer Science+Business Media Singapore 2016
S.S. Muthu and M.A. Gardetti (eds.), *Sustainable Fibres for Fashion Industry*,
Environmental Footprints and Eco-design of Products and Processes,
DOI 10.1007/978-981-10-0566-4_3

finishing—for the manufacture of different end products. Synthetic fibres, in particular, are dyed with different classes of structurally diverse dyes—such as acidic, basic, disperse, azo, diazo, anthraquinone and metal complex [3]. Unfortunately, the production of synthetic fibres from petrochemical sources is a global environmental problem. Nowadays, environmental protection and the search for alternative environmentally friendly fibres for use in textile processing have become a challenge for textile and polymer chemists [3, 4]. The demand for use of environmentally friendly fibres has been growing rapidly worldwide due to increased awareness of the environment, ecology and pollution control. Natural fibres from a wide range of plants, animals and minerals are important raw materials for various textile and non-textile applications [5–7]. Over the last few decades natural fibres have gained much attention since they manifest attractive features such as renewability and biodegradability [3, 8]. The sources of different fibres are shown in Fig. 1.

The domestication of animals and cultivation of plants for natural fibre production has a long history. Currently, a number of natural fibres are used to form fabrics for application in different fields and have revolutionized the textile industry. The characteristics and important properties of different natural fibres are discussed below.

1.1 Wool

Wool is an important million dollar natural fibre obtained from different animals, but mainly from sheep. It is complex in structure and essentially composed of three tissues: cuticle, cortex and medulla [9]. A schematic diagram of fine wool fibre is shown in Fig. 2.

The cuticle is the outermost surface of wool formed of thin cells and is mainly responsible for important properties such as wettability, felting and tactile

Fig. 1 Different sources of fibres

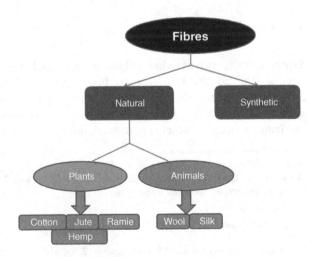

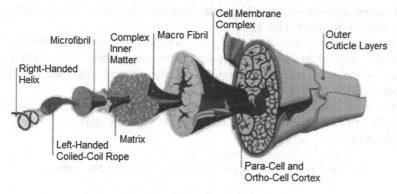

Fig. 2 Schematic diagram of the morphological components of a fine wool fibre [10]

behaviour. The cuticle is further composed of two distinct layers: the outermost layer is known as the "exo-cuticle" and beneath it is the innermost layer known as the "endo-cuticle" [11]. The exo-cuticle and endo-cuticle differ mainly in their cysteine proportions. High cysteine content in the exo-cuticle is responsible for high resistance to biological and chemical attacks. The endo-cuticle, on the other hand, is somewhat less resistant [12].

The cortex contributes about 90 % of wool fibre. The cortex consists of long closely packed cells. The tensile strength, elastic properties and the natural colour of wool are determined mainly by the nature of cortical cells. The medulla of wool fibre is a third type of cell which sometimes is a hollow canal [13].

Raw wool may contain many impurities: wool fat, perspiration products, dirt and vegetable matter such as burs and seeds. Wool fat is a yellowish wax-like substance, derived from fatty acids and a complex monohydric alcohol—cholesterol ($C_{27}H_{45}OH$)—or isocholesterol (its isomer). Suint (dried perspiration of sheep) is mainly composed of potassium salts of fatty acids, such as oleic ($C_{18}H_{34}O_2$) and stearic acids ($C_{18}H_{36}O_2$) plus some sulphate, phosphate and nitrogenous materials. Dirt includes mineral soil, wind-blown dust, faecal matter, skin flakes, discarded cuticle cells and fragments of fibres broken from photo-oxidized brittle tips. Scouring is a process developed to remove fat, suint and dirt from wool [14–16].

Wool is mainly composed of keratinous proteins, but 17 % of wool is non-keratinous. The remaining non-protein material consists of waxy lipids and a small amount of polysaccharides [17].

The building blocks of proteins consist of about 20 amino acids, which have the general formula:

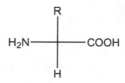

All amino acids found in wool have this structure, differing only in the side chain designated the R-group (Fig. 3).

When the amino group of one molecule is condensed with the carboxylic acid group of a second molecule a dipeptide is formed. Further condensation with another amino acid gives a tripeptide, and the process continues to form a polypeptide [13, 18, 19].

The general structure of a wool polypeptide is shown schematically in Fig. 4 where R represents amino acid side chains such as glycine, alanine, and phenylalanine.

$$R = H\text{-(glycine)}, CH_3\text{-(alanine)}, C_6H_5CH_2\text{-(phenylalanine)}$$

The individual polypeptide chains in wool are bonded together by various types of covalent crosslinks—mainly disulphide isopeptide crosslinks and noncovalent physical interactions such as hydrogen bonding, salt linkages or ionic and hydrophobic interactions (Fig. 5). Another type of covalent crosslink identified in wool is the isopeptide bond formed between amino acids containing acidic or basic groups [13, 20].

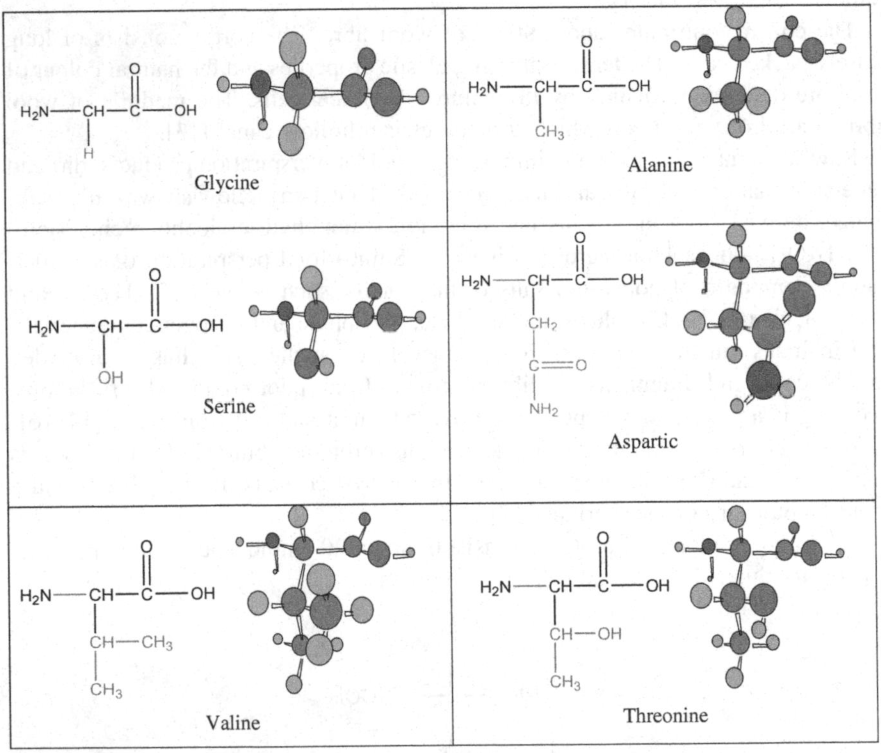

Fig. 3 Amino Acids present in wool

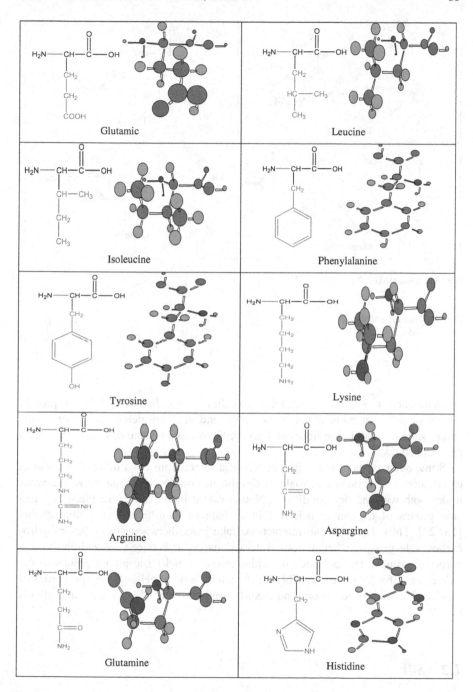

Fig. 3 (continued)

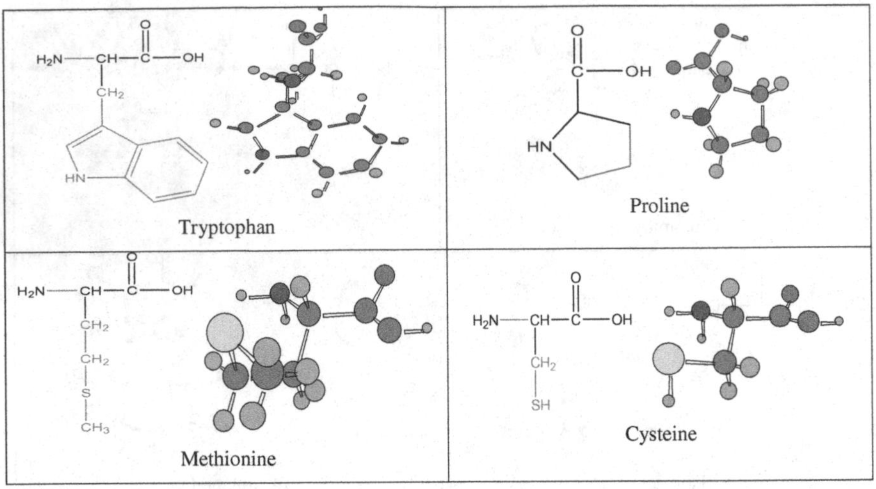

Fig. 3 (continued)

Fig. 4 General structure of wool peptide

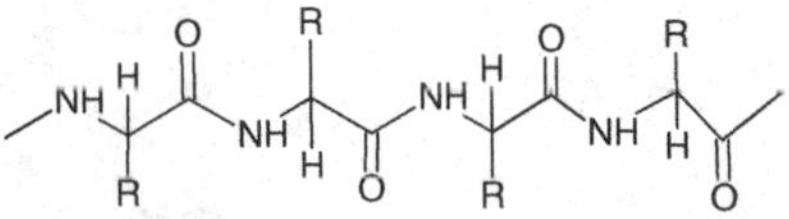

Disulphide crosslinks or cysteine crosslinks arise between different protein chains or different parts in the same chain and are considered the major bonds responsible for stabilizing fibres and giving them more resistance to chemicals and physical attack.

Some other types of secondary or physical interactions exist in wool in addition to covalent interactions and make a significant contribution to stabilize the wool under both wet and dry conditions. Non-covalent interactions take place between side groups of the amino acid within a chain or in different wool fibre chains [13, 21]. Thus, hydrophobic interactions take place between non-polar or hydrocarbon side groups, whereas ionic interactions occur between ionized amino and carboxyl groups. The presence of acidic carboxyl and basic amino groups in the side chains of wool are responsible for the amphoteric or pH-buffering properties of wool and its ability to absorb and desorb large amounts of both acids and alkalis [22, 23]

1.2 Silk

Silk is another common natural protein fibre produced by several insects; however, only the silk obtained from the cocoons of the larvae of *Bombyx mori* has been used

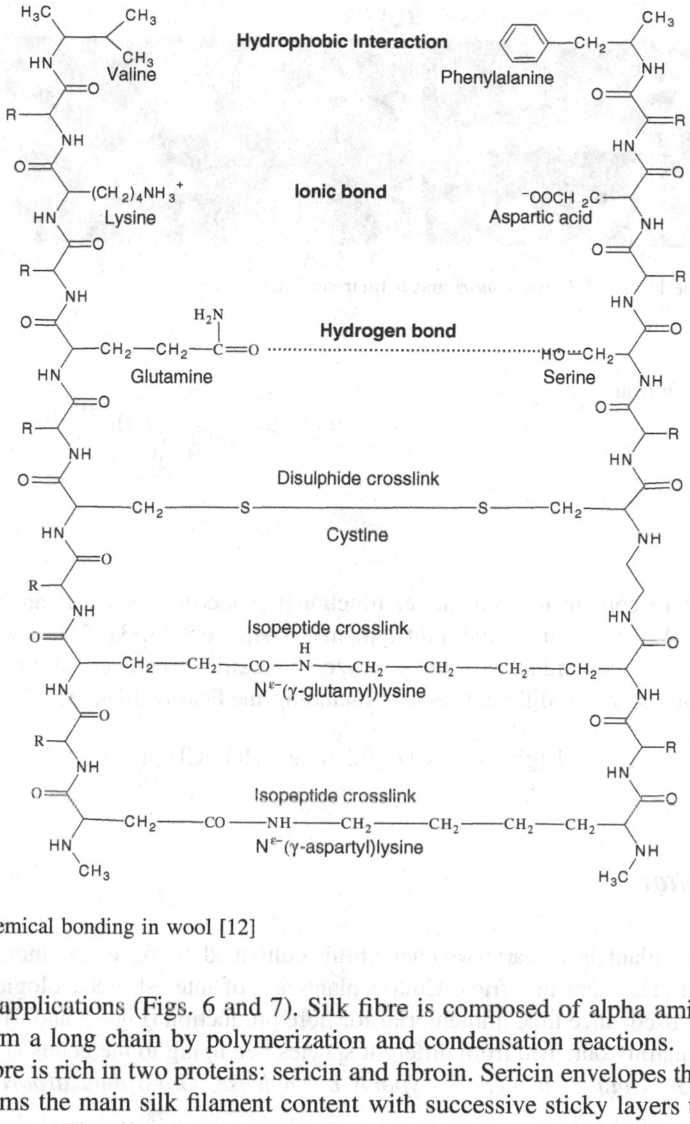

Fig. 5 Chemical bonding in wool [12]

in textile applications (Figs. 6 and 7). Silk fibre is composed of alpha amino acids which form a long chain by polymerization and condensation reactions.

Silk fibre is rich in two proteins: sericin and fibroin. Sericin envelopes the fibroin which forms the main silk filament content with successive sticky layers that help form a cocoon. The silk structure is simpler and similar to wool but contains amino acids having smaller pendant groups than those found in wool, allowing a pleated sheet structure rather than a helical one to occur. Silk has been extensively dyed with synthetic colourants. However, in view of their toxic nature natural colourants are extensively used by researchers nowadays for silk dyeing [24–27]. *Terminalia arjuna*, *Punica granatum* and *Rheum emodi* colourants have been applied to enzyme-pretreated silk for the development of softer options for silk dyeing [28]. To understand the natural dyeing mechanism in silk a few researchers have conducted thermodynamic and kinetic experiments [29, 30]. Natural dyes have been

Fig. 6 **a** The larvae of *Bombyx mori* and **b** their cocoons

Fig. 7 Silk structure

discovered to contain inherent novel functional properties—such as antibacterial, antifungal, UV blocking, deodorant and insect repellent [3, 31–34]. These properties of natural colourants have been utilized by textile researchers to functionalize silk for application in different sectors including medical textiles [35–38].

$$R = H\text{-(glycine)}, CH_3\text{-(alanine)}, HO-CH_2\text{-(serine)}$$

1.3 Cotton

The cotton plant is a warm-weather shrub cultivated in America, India, China, Egypt and elsewhere in Africa. Cotton plants are of interest archaeologically and have been used since time immemorial for fibre production (Figs. 8 and 9) [39, 40]. Cotton is mainly obtained from different species belonging to the genus *Gossypium* such as *Gossypium hirsute*, *Gossypium barbadense*, *Gossypium arboretum* and

Fig. 8 Cotton plant

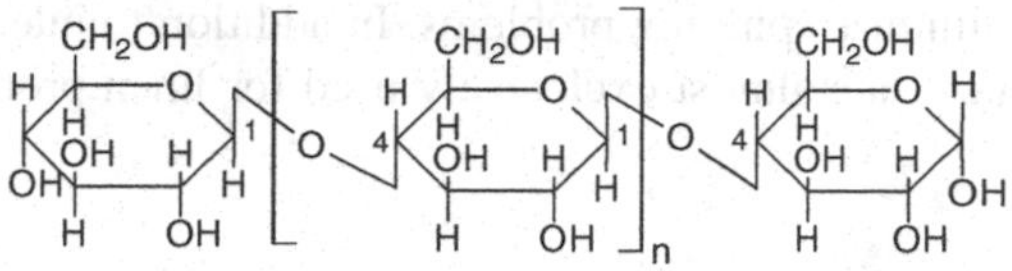

Fig. 9 Cellulose

Gossypium arboretum [41]. The general structure of cotton is more than 90 % cellulose, with the remainder being waxes, monomeric and polymeric sugars, residual protoplasm and minerals [42, 43].

Cotton fabrics were used extensively for tentage, tarpaulins and truck covers. However, with the discovery of microbe-resistant synthetic fibres—such as nylons, acrylics and polyesters—the use of cotton fabrics decreased to a large extent [44]. The textile industry uses several synthetic chemicals including synthetic dyes, salts and other additives for the manufacture of different products [45]. The cotton-dyeing industry is facing severe problems due to growing environmental awareness among different government and industrial firms. The natural colourants obtained from different plant, animal and mineral sources are nowadays used to counteract environmental pollution created by synthetic agents. Several recent reports are available on the dyeing and fastness properties of cotton dyed with natural agents [24, 32, 46–49]. Moreover, scientists have discovered that cotton can be imparted with multifunctional properties using different pre- and post-treatment methods [39, 50–54].

2 Bast Fibres

Bast fibres are potentially sustainable fibres obtained from the outer layers of the stems of various plants. They mainly consist of cellulose and varying amounts of lignin. The most important bast fibre crops attracting interest for their use in different sectors including the textile and construction industries are flax, ramie, jute and hemp. These fibres provide structural support to plants [55, 56].

2.1 Jute

Jute is a long, soft, shiny vegetable fibre predominantly cultivated in the Indian subcontinent, Bangladesh, India, China and Thailand and is spun into strong threads. It is a lignocellulosic fibre and is commonly known as "golden fibre". It is attracting more and more attention due to its outstanding properties—mainly its high moisture adsorbing ability [57, 58]. It has been widely utilized in the manufacture of flexible packing fabrics, carpet backing and decorative fabrics as well as in geo-textiles [59]. Jute is stiffer and harsher than cotton, hence chemicals and processes are required to minimize spinning problems. In addition to jute, flax fibres—particularly long-fibre flax—are almost exclusively used for linen production [60].

2.2 Ramie

Ramie is commonly known as "China grass", "white ramie", "green ramie" and "rhea". It is a herbaceous perennial flowering plant belonging to the family Urticaceae and is one of the oldest vegetable fibres—it was used to wrap mummies in Egypt. It is native to China and is increasingly being employed to make industrial sewing thread, packing materials, fishing nets and filter cloths [61, 62]. The main ramie-producing countries today are China, Brazil, India, South Korea and Thailand. It has been used to produce ropes, fishing nets, tents and tarpaulins as well as in fabrics for clothing and household furnishings (upholstery, canvas) [63]. To increase its use as a general textile fibre, ethylenediamine as a chelate molecule has been incorporated in the structure of raw ramie fibre using the crosslinking agent epichlorohydrin. The modified fibre has characters similar to that of wool fibre and a significantly improved dye uptake [64].

3 Natural Fibres from Waste Materials

Extensive efforts have been made over the past few years to use agricultural residues—generated in millions of tonnes every year from major crops, such as corn, wheat, rice, soybean, sorghum and sugarcane—or waste materials such as feather keratin and wheat gluten as raw materials for the production of cellulose and keratin fibres, respectively [45, 65, 66]. Corn stalks, corn husks, rice and wheat straw, rice husk, sorghum stalk and leaves, banana leaves, pineapple leaves, sugarcane stalks, hop stems, soybean straw, lotus petioles, *Cordia dichotoma* branches and sugarcane straw have been investigated as potential sources for cellulose fibres [44, 45]. Various methods—such as using bacteria and fungi—as well as mechanical and chemical methods are currently available for the extraction of natural cellulose fibres from waste products. Rosa et al. [67] isolated ultrathin cellulose nano-whiskers with diameters as fine as 5 nm from coconut husk fibres using acid hydrolysis. They used several techniques such as Fourier transform infrared spectroscopy (FTIR), transmission electron microscopy (TEM), thermogravimetric analysis (TGA) and X-ray diffraction to characterize cellulose nanowiskers and suggested that coconut husks could be a novel and renewable source for cellulose nanowhiskers. Likewise, Pasquini et al. [68] employed TEM and X-ray diffraction techniques to isolate cellulose nanowhiskers from *Cassava bagasse*. They also used sulphuric acid hydrolysis to extract cellulose—a by-product of starch industrialization—from *Cassava bagasse*. Zuluaga et al. [69] used a number of alkaline treatments—such as peroxide alkaline, peroxide alkaline–hydrochloric acid or 5 wt

% potassium hydroxide—to isolate cellulose microfibrils from vascular bundles of banana rachis. Both alkali treatment at 130 °C and subsequently sulphuric acid hydrolysis were employed by Li et al. [70] for the extraction of cellulose nano-whiskers from the branch bark of mulberry (*Morus alba* L.). Since mulberry bark is an agroindustrial waste, it therefore seems to be a readily available and renewable substitute for cellulose fibre. Alemdar and Sain [71] used the chemomechanical technique to isolate cellulose nanowhiskers from wheat straw and soy hulls and found that extracted cellulose had diameters in the range of 10–80 and 20–120 nm, respectively, from both these agrowastes.

Likewise feather keratin and wheat gluten which are annually renewable contain useful amino acids for fibre making and are an abundantly available cheap source for protein fibres [72]. This approach to fibre production from waste materials will help the fibre industry to be sustainable and add value to agrowastes.

4 Conclusion

In general, due to the relatively environmentally friendly, biodegradable and bio-compatible nature of natural fibres compared with their synthetic counterparts, they have been exploited as an attractive ecofriendly alternative for textile and other non-textile applications. The use of natural fibres for textile products could lead in future to the widespread production of functional textiles with improved properties for different end uses. Natural fibres have witnessed huge interest for use not only in the textile industry but also in other advanced fields (e.g., the construction, medical, environmental and safety technology sectors and, consequently, medical textiles). From the viewpoint of sustainability, fibres from animals, plants and agroindustrial wastes are potential alternatives to toxic synthetic fibres for use in the near future.

Acknowledgments Shahid-ul-Islam is very grateful to the University Grants Commission, Government of India, for financial support provided through the BSR Research Fellowship in Science for Meritorious Students.

References

1. Adeel S, Bhatti IA, Kausar A, Osman E (2012) Influence of UV radiations on the extraction and dyeing of cotton fabric with *Curcuma longa* L. Indian J Fibre Text Res 37:87–90
2. Islam S, Shahid M, Mohammad F (2013) Green chemistry approaches to develop antimicrobial textiles based on sustainable biopolymers-A review. Ind Eng Chem Res 52:5245–5260
3. Rather L, Shahid-ul-Islam, Mohammad F (2015) Study on the application of Acacia nilotica natural dye to wool using fluorescence and FT-IR spectroscopy. Fibers Polym 16:1497–505
4. Shahid M, Islam S, Mohammad F (2013) Recent advancements in natural dye applications: a review. J Clean Prod 53:310–331

5. Yusuf M, Ahmad A, Shahid M, Khan MI, Khan SA, Manzoor N, Mohammad F (2012) Assessment of colorimetric, antibacterial and antifungal properties of woollen yarn dyed with the extract of the leaves of henna (*Lawsonia inermis*). J Clean Prod 27:42–50
6. Islam S, Rather LJ, Shahid M, Khan MA, Mohammad F (2014) Study the effect of ammonia post-treatment on color characteristics of annatto-dyed textile substrate using reflectance spectrophotometery. Ind Crop Prod 59:337–342
7. Islam S, Mohammad F (2015) Natural colorants in the presence of anchors so-called mordants as promising coloring and antimicrobial agents for textile materials. ACS Sustain Chem Eng 3:2361–2375
8. Babu KM (2015) Natural textile fibres: Animal and silk fibres. In: Sinclair R (ed) Textiles and fashion, Woodhead Publishing, pp 57–78
9. Silva CJSM, Prabaharan M, Gübitz G, Cavaco-Paulo A (2005) Treatment of wool fibres with subtilisin and subtilisin-PEG. Enzym Microb Technol 36:917–922
10. Feughelman M (1997) Introduction to the physical properties of wool, hair and other—keratin fibres. In: Mechanical properties and structure of alpha-keratin fibres: wool, human hair and related fibres, UNSW Press, pp 1–14
11. Bradbury JH (1973) In: Anfinsen Jr CB, Edsall JT, Richards FM (eds) Advances in protein chemistry, vol 27, Academic Press, New York, p 111
12. Lindley H (1977) In: Asquith, RS (ed) Chemistry of natural protein fibers, Plenum Press, New York, p 147
13. Rippon JA (2013) The structure of wool. Wool dyeing, pp 1–51
14. Broughton RM Jr, Brady PH, Hall DM, Adanur S (1995) In: Adanur S (ed) Wellington sears handbook of industrial textiles. CRC Press Inc, Pennsylvania
15. Trotman ER (1984) Dyeing and chemical technology of textile fibres. Wiley, New York
16. Cook JG (1984) Handbook of textile fibres: natural fibres. Woodhead Publishing Limited, Abington
17. Lewis DM (1992) Dyers so, colourists. Wool dyeing. Society of Dyers and Colourers, Bradford
18. Ziegler K (1977) In: ed. Asquith RS (ed) Chemistry of natural protein fibers. Plenum Press, New York, p 267
19. Lungren HP, Ward WH (1963) In: Borasky R (ed) Ultrastructure of protein fibers, Academic Press, New York
20. Christoe JR, Bateup BO (1987) Wool Sci Rev 63:25
21. Baumann H (1979) In: Parry DAD, Creamer LK (eds) Fibrous proteins: scientific, industrial and medical aspects, vol 1, Academic Press, London, p 299
22. Rigby B, Robinson MS, Mitchell TW (1982) J Text Inst 73:94
23. Gillespie JM, Darskus RL (1971) Aust J Bio Sci 24:1189
24. Chairat M, Bremner JB, Chantrapromma K (2007) Dyeing of cotton and silk yarn with the extracted dye from the fruit hulls of mangosteen, Garcinia mangostana linn. Fibers Polym 8:613–619
25. Dasa D, Maulik SR (2007) Dyeing of wool and silk with *Bixa orellana*. Indian J Fibre Text Res 32:366–372
26. Dasa D, Maulik SR, Bhattacharya SC Colouration of wool and silk with *Rheum emodi*. Indian J Fibre Text Res 33:163–170
27. Jung Y, Bae D (2014) Natural dyeing with black cowpea seed coat. I. Dyeing properties of cotton and silk fabrics. Fibers Polym 15:138–144
28. Vankar PS, Shanker R, Verma A (2007) Enzymatic natural dyeing of cotton and silk fabrics without metal mordants. J Clean Prod 15:1441–1450
29. Chairat M, Rattanaphani S, Bremner JB, Rattanaphani V (2005) An adsorption and kinetic study of lac dyeing on silk. Dyes Pigm 64:231–241
30. Kongkachuichay P, Shitangkoon A, Chinwongamorn N (2002) Thermodynamics of adsorption of laccaic acid on silk. Dyes Pigm 53:179–185
31. Ali NF, El-Mohamedy RSR (2011) Eco-friendly and protective natural dye from red prickly pear (*Opuntia Lasiacantha Pfeiffer*) plant. J Saudi Chem Soc 15:257–261

32. Ali S, Hussain T, Nawaz R (2009) Optimization of alkaline extraction of natural dye from Henna leaves and its dyeing on cotton by exhaust method. J Clean Prod 17:61–66
33. Baliarsingh S, Behera PC, Jena J, Das T, Das NB (2015) UV reflectance attributed direct correlation to colour strength and absorbance of natural dyed yarn with respect to mordant use and their potential antimicrobial efficacy. J Clean Prod 102:485–492
34. Bechtold T, Mahmud-Ali A, Mussak RAM (2007) Reuse of ash-tree (*Fraxinus excelsior* L.) bark as natural dyes for textile dyeing: process conditions and process stability. Color Technol 123:271–279
35. Baliarsingh S, Panda AK, Jena J, Das T, Das NB (2012) Exploring sustainable technique on natural dye extraction from native plants for textile: identification of colourants, colourimetric analysis of dyed yarns and their antimicrobial evaluation. J Clean Prod 37:257–264
36. Ajmal M, Adeel S, Azeem M, Zuber M, Akhtar N, Iqbal N (2014) Modulation of pomegranate peel colourant characteristics for textile dyeing using high energy radiations. Ind Crop Prod 58:188–193
37. Lee Y-H, Hwang E-K, Baek Y-M, Kim H-D (2014) Deodorizing function and antibacterial activity of fabrics dyed with gallnut (*Galla Chinensis*) extract. Text Res J 85:1045–1054
38. Lee Y-H, Hwang E-K, Kim H-D (2009) Colorimetric assay and antibacterial activity of cotton, silk, and wool fabrics dyed with peony, pomegranate, clove, coptis chinenis and gallnut extracts. Materials 2:10–21
39. Abdel-Mohdy F, Fouda MMG, Rehan M, Aly A (2008) Repellency of controlled-release treated cotton fabrics based on cypermethrin and prallethrin. Carbohydr Polym 73:92–97
40. Tutak M, Korkmaz NE Environmentally friendly natural dyeing of organic cotton. J Nat Fibers 9:51–59
41. Kirk RE, Othmer DF, Mark HF (1965) Encyclopedia of chemical technology, Interscience Publishers
42. Fassihi A, Hunter L (2015) Application of an automatic yarn dismantler to track changes in cotton fiber properties during processing on a miniature spinning line. J Nat Fibers 12:121–131
43. Eser F, Onal A (2015) Dyeing of wool and cotton with extract of the nettle (*Urtica dioica* L.) leaves. J Nat Fibers 12:222–2231
44. Islam S, Shahid M, Mohammad F (2013) Perspectives for natural product based agents derived from industrial plants in textile applications—a review. J Clean Prod 57:2–18
45. Islam S, Mohammad F (2014) Emerging green technologies and environment friendly products for sustainable textiles. In: Muthu SS (ed) Roadmap to Sustainable Textiles and Clothing. Springer, Singapore, pp 63–82
46. Adeel S, Ali S, Bhatti IA, Zsila F (2009) Dyeing of cotton fabric using pomegranate (Punica granatum) aqueous extract. Asian J Chem 21:3493–3499
47. Grifoni D, Bacci L, Di Lonardo S, Pinelli P, Scardigli A, Camilli F (2014) UV protective properties of cotton and flax fabrics dyed with multifunctional plant extracts. Dyes Pigm 105:89–96
48. Davulcu A, Benli H, Şen Y, Bahtiyari Mİ (2014) Dyeing of cotton with thyme and pomegranate peel. Cellulose 21:4671–4680
49. Vankar PS, Shanker R, Dixit S, Mahanta D, Tiwari SC (2008) Sonicator dyeing of modified cotton, wool and silk with Mahonia napaulensis DC. and identification of the colorant in Mahonia. Ind Crop Prod 27:371–379
50. Abdel-Halim ES, Abdel-Mohdy FA, Fouda MMG, El-Sawy SM, Hamdy IA, Al-Deyab SS (2011) Antimicrobial activity of monochlorotriazinyl-β-cyclodextrin/chlorohexidin diacetate finished cotton fabrics. Carbohydr Polym 86:1389–1394
51. Adeel S, Bhatti IA, Kausar A, Osman E (2012) Influence of UV radiations on the extraction and dyeing of cotton fabric with Curcuma longa L. Indian J Fibre Text Res 37:87–90
52. Chung YS, Lee KK, Kim JW (1998) Durable press and antimicrobial finishing of cotton fabrics with a citric acid and chitosan treatment. Text Res J 68:772–775
53. Koh E, Hong KH (2014) Gallnut extract-treated wool and cotton for developing green functional textiles. Dyes Pigm 103:222–227

54. El-Tahlawy KF, El-Bendary MA, Elhendawy AG, Hudson SM (2005) The antimicrobial activity of cotton fabrics treated with different crosslinking agents and chitosan. Carbohydr Polym 60:421–430
55. Das B, Chakrabarti K, Tripathi S, Chakraborty A (2014) Review of some factors influencing jute fiber quality. J Nat Fibers 11:268–2681
56. Pan NC, Chattopadhyay SN, Day A (2007) Dyeing of biotreated jute fabric. J Nat Fibers 4:67–76
57. Sreenath HK, Shah AB, Yang VW, Gharia MM, Jeffries TW (1996) Enzymatic polishing of jute/cotton blended fabrics. J Ferment Bioeng 81:18–20
58. Liu L, Wang Q, Xia Z, Yu J, Cheng L (2010) Mechanical modification of degummed jute fibre for high value textile end uses. Ind Crops Prod 31:43–47
59. Kicińska-Jakubowska A, Bogacz E, Zimniewska M (2012) Review of natural fibers. Part I—vegetable fibers. J Nat Fibers 9:150–167
60. Kymäläinen H-R, Sjöberg A-M (2008) Flax and hemp fibres as raw materials for thermal insulations. Build Environ 43:1261–1269
61. Angelini LG, Lazzeri A, Levita G, Fontanelli D, Bozzi C (2000) Ramie (Boehmeria nivea (L.) Gaud.) and Spanish Broom (Spartium junceum L.) fibres for composite materials: agronomical aspects, morphology and mechanical properties. Ind Crops Prod 11:145–161
62. Brühlmann F, Leupin M, Erismann KH, Fiechter A (2000) Enzymatic degumming of ramie bast fibers. J Biotechnol 76:43–50
63. Pandey SN (2007) Ramie fibre: part I. Chemical composition and chemical properties. A critical review of recent developments. Text Prog 39:1–66
64. Liu Z-T, Yang Y, Zhang L, Sun P, Liu Z-W, Lu J (2008) Study on the performance of ramie fiber modified with ethylenediamine. Carbohydr Polym 71:18–25
65. Islam S, Mohammad F (2015) High-energy radiation induced sustainable coloration and functional finishing of textile materials. Ind Eng Chem Res 54:3727–3745
66. Islam S, Shahid M, Mohammad F (2014) Future prospects of phytosynthesized transition metal nanoparticles as novel functional agents for textiles. In: Syväjärvi AT (ed) Advanced materials for agriculture, food, and environmental safety. Wiley, pp 265–90
67. Rosa MF, Medeiros ES, Malmonge JA, Gregorski KS, Wood DF, Mattoso LHC, Imam SH (2010) Cellulose nanowhiskers from coconut husk fibers: Effect of preparation conditions on their thermal and morphological behavior. Carbohydr Polym 81:83–92
68. Pasquini D, Morais Teixeira E, Silva Curvelo AA, Belgacem MN, Dufresne A (2010) Extraction of cellulose whiskers from cassava bagasse and their applications as reinforcing agent in natural rubber. Ind Crops Prod 32:486–490
69. Zuluaga R, Putaux JL, Cruz J, Vélez J, Mondragon I, Gañán P (2009) Cellulose microfibrils from banana rachis: Effect of alkaline treatments on structural and morphological features. Carbohydr Polym 76:51–59
70. Li R, Fei J, Cai Y, Li Y, Feng J, Yao J (2009) Cellulose whiskers extracted from mulberry: a novel biomass production. Carbohydr Polym 76:94–99
71. Alemdar A, Sain M (2008) Isolation and characterization of nanofibers from agricultural residues–Wheat straw and soy hulls. Bioresour Technol 99:1664–1671
72. Poole AJ, Church JS, Huson MG (2008) Environmentally sustainable fibers from regenerated protein. Biomacromolecules 10:1–8

Sabai Grass: Possibility of Becoming a Potential Textile

Asimananda Khandual and Sanjay Sahu

Abstract Clothing made up of natural fibres has prompted growing interest among users over time. It is environmentally friendly and benign from a health perspective especially when compared with synthetic materials and those natural fibres that are prone to degrade. Besides the development of routine natural fibres such as cotton, hemp, silk and wool, research has been directed towards developing technologies and optimizing new ways of utilizing some unconventional plant fibres. These plant fibres not only enrich the textile materials domain, but also bring certain economic and social benefits. Sabai grass (*Eulaliapsis binata*), a perennial plant belonging to the family Poaceae is grown in many Asian countries such as China, India, Pakistan, Nepal, Bhutan, Myanmar, Thailand, Malaysia and the Philippines. Since it's thin and long leaves possess high-quality fibre, it is used as a major raw material for the paper industry as it is superior to most other available grasses. In India, it is being used for paper making since 1870. For their flexibility and strength, the leaves are utilized for making ropes and other rope-based utility items. Most importantly, sabai grass has a prominent role to play in the sustainability of tribal economics of some regions of various countries. The very survival of tribal people completely depends on sabai grass processing in some parts of India. The components of *Eulaliopsis binata* are reported to have cellulose contents up to 52 %, which is more than sisal and palm. The fundamental characteristic of this fiber is that it is better with a lignin content close to 18.5 %. At present (as of 2016) the fibre and the way it is processed has not been fully developed, because it is mainly used to make paper, conserve soil and fabricate yarn for knitting. It is used in making of many tribal products, craft items and pulp. Some recent research, aimed at developing degumming and bleaching in a single bath using an ultrasonic-assisted process, concluded that the fiber deserves to be developed further as a means of deriving cellulose. It holds a lot of promise both for apparel and

A. Khandual (✉)
Fashion & Apparel Technology, College of Engineering & Technology (CET), Bhubaneswar 751003, Odisha, India
e-mail: asimte@cet.edu.in

S. Sahu
Clearity Specialties LLP, Thane, India

© Springer Science+Business Media Singapore 2016   45
S.S. Muthu and M.A. Gardetti (eds.), *Sustainable Fibres for Fashion Industry*,
Environmental Footprints and Eco-design of Products and Processes,
DOI 10.1007/978-981-10-0566-4_4

technical textiles. This chapter will discuss the history and socioeconomic importance, chemical constitution, fibre properties, recent research and potential
applications.

1 Introduction

Textiles have been a fundamental part of human civilization since inception and
natural fibres have been cultivated and processed from various plants and animals
since antiquity. The livelihood of millions of people depend on natural fibre production and processing in the face of relentless competition from synthetic counterparts over the past half-century in clothing, household furnishings, industry and
agriculture. The success of synthetics has been mainly due to cost and customised
applications. After World War II, the buildup of synthetic fibres significantly
decreased the use of natural fibres. With continuous increase in petrochemical
prices (despite the current lull) and environmental considerations, there has been a
revival of natural fibre uses in the textile, industrial fibre, plastics and automotive
industries. There are many reasons for the current turnaround. The most important
revolve around choices are as follows [1]:

(a) Responsible choice—natural fibres are of major economic importance to many
    developing countries. They are vital to the livelihoods and food security of
    millions of small-scale farmers and processors. By choosing natural fibres, we
    boost the sector's contribution to economic growth and helps fighting hunger
    and rural poverty.
(b) Healthy choice—natural fibres provide natural ventilation. Cotton feels com-
    fortable on a hot day. Wool acts as an insulator against both cold and heat.
    Coconut fibres have natural resistance to fungi and mites. Hemp fibre has
    antibacterial properties and linen is the most hygienic textile for hospital bed
    sheets. Considering the health hazards associated with the use of fibres like
    asbestos, glass and carbon, opting for renewable resources in the form of
    natural fibres makes a lot of sense.
(c) Sustainable choice—natural fibres are not only important to producers/industry,
    but also to consumers and the environment. There is a global thrust towards the
    green concept: renewable feed stocks in polymer products, bio-degradability,
    reduction of carbon emissions and recyclable materials that are cost-effective in
    terms of energy.
(d) Technical choice—many natural fibres have shown good mechanical strength,
    reduced weight and reasonably lower cost. They can be reinforced with
    thermoplastic panels for high-tech application. Fibres that give strength and
    stability to plants are being incorporated in an ever widening range of
    industrial products.
(e) Fashionable choice—eco-fashion conceptualizes natural fibres and the gar-
    ments made from them as being sustainable at every stage of their lifecycles.

Fig. 1 Comparing bio-fibers with inorganic and organic fibres [3, 4]

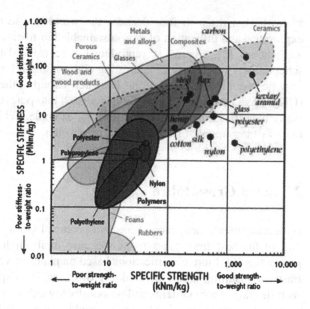

Table 1 Bio-fibres in car manufacture [4, 5]

Auto makers	Bio-fibres	Auto parts	Model
Daimler Chrysler	• Flax • Hemp • Sisal • Coconut • Caoutchouc • Abaca • Castor oil seed	• Door cladding • Seatback linings • Package shelves • Seat bottom • Back cushions • Head restraints • Under-floor body panels • Flexible tubing for fuel • Brake systems	Mercedes Benz Class A, C, E, S
BMW	• Flax • Sisal • Cotton • Wool • Wood fibre	• Interior door linings and panels • Soundproofing • Upholstery • Seatback cushions	3, 5, 7 series
Toyota	• Kenaf	• Package shelves • Body structure	Lexus i-foot/i-unit
GM	• Kenaf/flax mixture • Wood fibre	• Package trays • Door panel inserts • Seatbacks • Cargo area floor	Satum, L300s Opel, Vectras Cadillac DeVille, GMC Envoy, Chevrolet TrailBlazer
Honda	• Wood fibre	• Cargo area floor	Pilot SUV
Ford	• Corn • Wood fibre	• Goodyear tyres • Sliding door inserts	Fiestas Ford, Freestar

In the last three to four decades the world has seen a growth in interest for exploration and utilization of new sustainable bio-fibers and composites for various applications [2]. Figure 1, reproduced here from data in the public domain, schematically represents an interesting comparison of bio-fibers with inorganic and organic fibers by an Engineering Research Group at the University of Cambridge (UK). There has been a great deal of research into potential applications of fibres in many industries not related to textiles. For example, Table 1 shows the trend of car manufacturer uptake of natural fibre composites.

2 Sabai Grass Fibre

As already mentioned, sabai grass is a perennial grass of the Poaceae family. It is one of the best fibre grass plants because of its high fiber quality and ease of production [6]. Fibre classifications based on plants or vegetables is given in Fig. 2. Ease of planting, good perennial growth, wide adaptability, stress resistance, well-developed root systems and propensity towards dense plant populations are the noteworthy attributes of sabai grass, making it a suitable species for soil, water conservation and wasteland construction [7, 8]. In addition, sabai grass's good flexibility and strength, high leaf fibre content (>55 %), low lignin content (<14 %), excellent average fibre length (20 mm) make it one of the best fibre grass plants for the paper industry and for rayon and woven materials [6, 9–11]. Large-scale cultivation of sabai grass on barren hills and slopes has already been practised and proved to have fast ecological benefits [12, 13]. Sabai grass's family characteristics show that it is a relative of cereal crops. It has a high frequency of autonomous apospory [14]. Apospory, sometimes called "apomixy", means the production of a sexual generation from an asexual generation without the intervention of spores. Sabai grass is seen in the wild in China [9, 15], India and Southeast Asia; growing

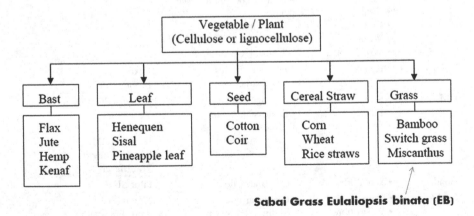

Fig. 2 Classification of vegetable/plant-based fibre

in different ecological habitats [16, 17]. Sabai grass is also popularly used as a construction material for thatches, walls, roofs and ropes, filler material in plastics and in matrix mud systems [18, 19].

3 Morphology of Sabai Grass Fibre

Certain unique morphological characteristics such as it open and loose anatomical features with low lignin content make sabai grass fibre easy to pulp at low temperatures and readily bleachable. The fibre dimensions and water drainage characteristics of pulps mean the fibre does not require extensive refining to develop fibrillation, hydration and interfibre bonding (Table 2). These make it suitable for special quality-grade paper. A typical image of the fibre from a scanning electron microscope is depicted in Fig. 3 [19].

The average fibre length of sabai grass is 2.4 mm; this compares with 17 mm of bamboo. Therefore, the mechanical strength properties of sabai grass are better than those of bamboo. Based on morphological characteristics (Table 3), sabai grass is comparable and even better in some characteristics than bamboo fibre. Microscopic examination reveals that sabai grass fibres are long, thin and taper off in some parenchymatous cells. Cell size varies from 3 to 20 μm. Cell width is 1.6–3.2 μm

Table 2 Other grass fiber dimensions [21, 22]	Property	Sugarcane bagasse	Bamboo	Esparto	Sabai
	Length (mm)	2.8	2.7–4	1.1	0.5–4.9
	Width (mm)	34.1	15	9	9–16

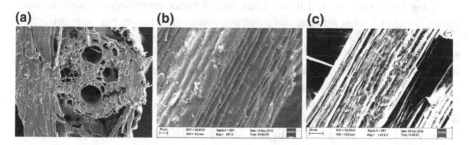

Fig. 3 Scanning electron microscope images **a** Sabai grass fibre **b** Sabai grass scoured Bleached SEM1 **c** Sabai grass SEM1

Table 3 Morphological characteristics of sabai grass and bamboo fibre [22, 23]

Particulars	Sabai grass
Colour	Brownish yellow
Fibre length L (mm)	2.4
Fibre width D (μm)	9.90
Lumen width d (μm)	5.75
Cell wall thickness w (μm)	2.11
Flexibility coefficient $d/D \times 100$	57.67
Slenderness ratio L/D	266
Rigidity coefficient $2w/D$	0.42
Wall fraction $(2w/D) \times 100$	42
Runkel ratio $2w/d$	0.73

and cell length 30–50 μm. The epidermal cells of sabai grass can be dissolved to some extent using alkalis [18, 22, 23].

4 Chemical Composition

Eulaliopsis binata exhibits the typical chemical composition of the grass family containing more ash and less lignin. It is an excellent natural cellulose material, noted for its long, strong and tough fibres. Sabai grass is rich in cellulose and has a low-lignin content. In addition, the ash content of *Eulaliopsis binata* is greater than that of bamboo. All lignocellulosic materials are generally composed of cellulose, hemicellulose and lignin. The high degree of polymerization and molecular weight of sabai grass indicates that it is a potential resource for producing high-intensity fibre products and bio-composites. Understanding the structure and properties of natural fibres can lead to a large number of applications [18].

Along with cellulose and lignin a number of minor components—such as wax, pectin, inorganic salts, nitrogenous substances and colouring matter—are present in sabai grass. Hemicellulose is non- homogeneous and generally comprises polysaccharides of relatively low molecular weight (i.e., hexoses such as galactose, mannose and pentoses such as xylose and uronic acid). In extracted long vegetable fibres, the content of cellulose varies from 81 % in pineapple to about 43 % in coir. Of these bast fibres, jute has the highest lignin content at about 14 % and ramie (a flowering plant of the nettle family) the lowest at 0.6 %. Coir, whose fibre consists of husk, contains about 45 % lignin. Momin (1950) was the first to study the chemical composition of sabai grass and, as a result, applications such as pulping and the production of printing paper have also been suggested [24]. Lignin is the compound that supports plant rigidity. It is a complex three-dimensional copolymer of aliphatic and aromatic constituents with a very high molecular weight and is most commonly derived from wood. It is an integral part of the cell walls in the plants [18, 22, 23]. The term was introduced in 1819 by de Candolle who derived it

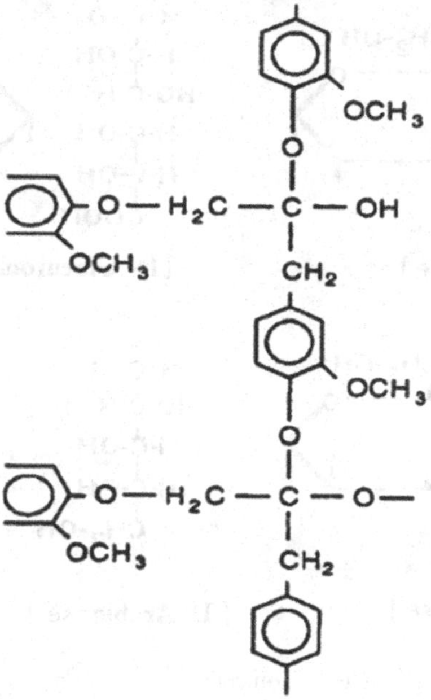

Fig. 4 Partial structure of lignin [25]

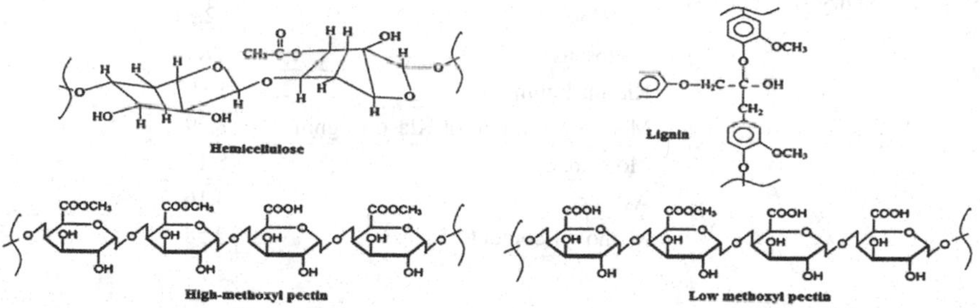

Fig. 5 Chemical structure of hemicellulose, lignin and pectins [25–27]

from the Latin word *lignum*, meaning wood. Lignin is amorphous and hydrophobic in nature (Figs. 4 and 5).

Unlike cellulose, which contains only a 1, 4-β-glucopyranose ring, hemicellulose contains different types of sugar units such as D-xylopyranose, D-glucopyranose, D-galactopyranose, L-arabinofuranose, D-mannopyranose and D-glucopyranosyluronic acid with a minor amount of other sugars [28, 29] (Fig. 6).

Chand and Rastogi (1992) studied the chemical composition of sabai grass fibre based on dry weight (Table 4) [29]. Chemical constituents such as cellulose, lignin and ash are different in different sabai grass samples.

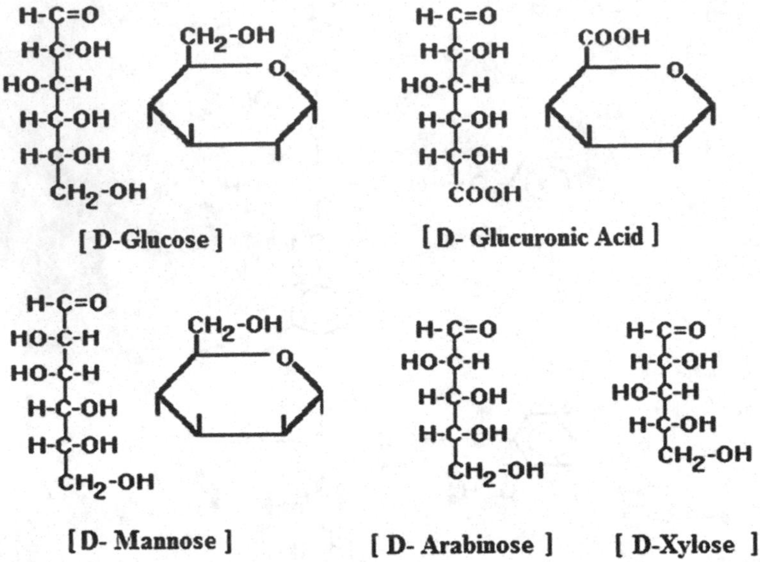

Fig. 6 Different types of sugar units in hemicellulose

Constituent	Percentage present
Cellulose	52.34
Pentosans	27.2
Klason lignin	16.07
Methoxyl content of Klason lignin	8.79
Moisture	8.12
Ash	4.16
Methoxyl content of grass	2.58

Table 4 Analysis of sabai grass fibre on dry weight basis [29]

Tang et al. [20] studied the chemical constituents of sabai grass—both untreated and acid hydrolysed with 0.5 % sulphuric acid in details. They found a small reduction in intrinsic viscosity, degree of polymerization and molecular weight in acid-hydrolysed sabai grass compared with the untreated one and concluded that dilute acid hydrolysis has little effect on cellulose degradation. However, hemicellulose was mostly dissolved, and lignin hemicellulose–cellulose interactions were also partially disrupted in this hydrolysis process. The cellulose in *Eulaliopsis binata* was identified as cellulose I with low crystallinity. The chemical composition is presented in Table 5.

Tyagi et al. (2004) described particulars of the solubility and content of *Eulaliopsis binata* during soda and soda–anthraquinone (AQ) pulping in great depth[32] (Table 6).

Table 5 Chemical constituents of sabai grass fibre untreated and acid hydrolysed [30]

Constituent	Percentage present	
	Untreated	Acid hydrolysed
Ash	5.77	0.76
Klason lignin	18.2	17.8
Glucose	49.9	45.7
Xylose	19.8	1.91
Arabinose	2.78	0.43
Galactose	0.94	0.11
Mannose	0.28	0.22
Glucuronide	0.15	Traces
Galacturonic acid	0.24	Traces
Weight	100 %	68.6 %

Table 6 Solubility and content of *Eulaliopsis binata* during soda and soda–AQ pulping

Particulars	(%)
Cold-water solubility	3.60
Hot-water solubility	9.50
Alcohol–benzene solubility (1:2 v/v)	4.10
1 % NaOH solubility	39.70
Lignin	22.00
Pentosan	23.90
α-Cellulose	49.00
Ash	6.00

5 Fibre Extraction and Analysis

Recently, Tian et al. [34] studied acid hydrolysis in greater depth followed by a single-bath degumming process to extract sabai grass fibres. They found that some tissue cells get destroyed during acid hydrolysis due to the solubilization of hemicelluloses and other minor constituents as well as to structural changes of lignin as a result of increased specific surface area of the *Eulaliopsis binata* fibre. Acid pretreatment was given using an H_2SO_4 solution (1 mL/L) at a temperature of 50 °C and a material-to-liquor ratio of 1:15 for 60 min. The sabai grass fibres were then treated in the bath degumming process in alkali–H_2O_2 [31] to extract *Eulaliopsis binata* fibres as per the recipe and conditions given in the following bullet list:

- Material-to-liquor ratio: 1:15
- NaOH solution: 5 g/L
- $MgSO_4 \cdot 7H_2O$ solution: 0.1 g/L
- H_2O_2 solution: 4 g/L
- Amino trimethylene phosphonic acid + magnesium chloride ($MgCl_2$) as H_2O_2 stabilizer: 1.2 g/L

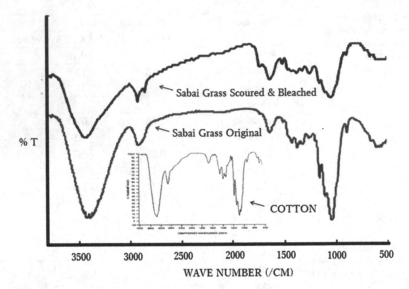

Fig. 7 FTIR of *Eulaliopsis binata* original fibres as well as scoured and bleached fibres (*lower graph line* is cotton)

- Temperature: 99 °C
- Time: heating in the water bath for 150 min
- 0.01 g of finely ground powder of the fibre mixed with 0.2 g of Fourier transform infrared (FTIR) grade KBr and mixed well under dry conditions.

Tablets were made by pellet-making apparatus supplied with the Perkin Elmer FTIR, and spectra were measured for untreated and extracted *Eulaliopsis binata* fibres. Figure 7 depicts FTIR spectra of treated and untreated sabai grass along with FTIR spectra of cotton fibre for ease of comparison. It can be stated that most cellulose structures in *Eulaliopsis binata* are not disrupted by acid hydrolysis and the fingerprint transmittance peaks of cellulose are preserved in all cases (peaks of 3402 cm^{-1} for the OH group; the 1431-, 1166-, 1059-cm^{-1} fingerprint areas can be attributed to the cellulose structure) [30].

The absence of a vibration peak in the $\cong$1735-cm^{-1} region [31], which dominantly contributes to CO stretching of methylester and carboxylic acid in the pectin or acetyl group in hemicelluloses, confirms that pectin and hemicelluloses have been eliminated or removed in the case of extracted fibres. So the process here may be regarded as a scouring and bleaching process as suggested by Gierlinger et al. [32]. Stewart et al. [33] regarded the absorption band in the 1510-cm^{-1} region as confirming that lignin is present in sabai grass raw fibre but found to be absent in extracted fiber; corroborating the fact that non-cellulose substances such as pectin, lignin, hemicellulose and wax are being removed or reduced and that the cellulose I structure is preserved. X-ray diffraction (XRD) of the extracted fibre (Fig. 8) also confirms the preserved cellulose I structure of the fibre with increased crystallinity and orientation compared with the raw fibre.

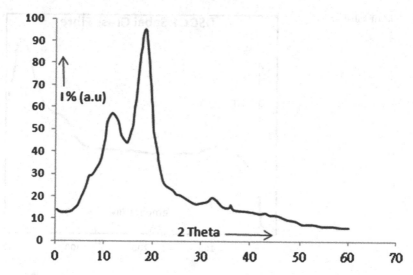

Fig. 8 XRD of hydrolysed *Eulaliopsis binata*

6 Physical Properties

Plant fibres of different age, source and place have different chemical compositions and hence different properties. It is worth mentioning here the work of Chand et al. [18, 19, 29] who studied the detailed mechanical and thermal properties of sabai grass fibre and their stress–strain curves. They subjected the fibre to differential thermal analysis (DTA) and differential scanning calorimetry (DSC). They also provide many other relevant details [18, 19, 29]. The stress–strain curves of sabai grass fibre are initially linear but followed by a non-linear portion. The ultimate tensile strength of sabai grass fibre is about 76 MPa. This is mainly due to the cellulosic content arranged at a microfibril angle with the fibre and bonded to each other by lignin, parallel to the fringed fibril structure. The crystalline regions are embedded in non-crystalline/amorphous regions, as is the case with many plant fibres. Thermo-gravimetric loss was initially found at 75 °C and continued up to 100 °C (Fig. 9). DSC analysis showed the endothermic peak at 108 °C.

The most significant loss of fibre starts at 244 °C and continues up to 520 °C in thermogravimetry [29]. Endothermic peaks were found at 320 and 380 °C in DSC—much like the case in derivative thermogravimetry (DTG) at 380 °C—implying pyrolysis had occurred at this DSC peak. Chand et al. [29] likened this phenomenon of sabai grass fibre to sisal fibre degradation [19, 29]. Sabai grass fibre is similar to other natural fibres and can be useful in making composites for housing. Table 7 (from Tian et al. [34]) is reproduced here to compare the properties of sabai grass fibre with commonly used natural fibres like cotton and flax. It is longer and coarser than flax and cotton, its moisture content is comparable with cotton, its stiffness or

Fig. 9 DSC of sabai grass
fibre

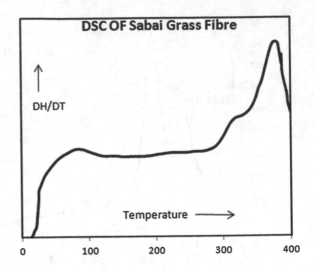

Table 7 Fibre properties of
sabai grass fibre compared
with those of cotton and flax
[34]

Fiber properties	Sabai grass fibre	Flax	Cotton
Strength (cN/dtex)	3.5–6.9	4.1–5.5	1.9–3.5
Elongation (%)	5.0–6.1	1.6–3.3	6.0–9.0
Length (mm)	63–70	17–25	15–56
Fineness (dtex)	3.7–4.3	1.7–3.3	1.5–2.0

tensile modulus is on a par with flax [34, 35]. As far as the most important properties
are concerned, it can be regarded as an alternative source to natural cellulosic fibre.

7 Antibacterial Properties

Experimental results from the antibacterial property study of sabai grass fibres by
Tian et al. [34] found the fibre to be better than cotton at *Escherichia coli* inhibition
and can be regarded to have moderate antibacterial properties. They concluded that
this intrinsic antimicrobial property may be attributed to compounds—such as
tannic acid, avones and phenols—that inhibit *E. coli* [36] and, second, the
degumming process reduces/eliminates pectin—a nutrient supporter of *E. coli*.

8 Some Key Issues and Scope

8.1 Fiber Extraction and Sustainability

Sabai thrives well in regions with annual rainfall of 30–60 in [37] (a number of Asian countries fit the bill), and the annual harvest is around November. After that the culms (stalks) are sun-dried on roads and pavements for 2–3 days when their natural greenish yellow colour develops. The natural colour of the grass also fades as a result of night dew or frost. The fibres are dried, brushed and baled for easy transportation. They are then stored in bundles and subsequently taken for rope preparation. The culms are then processed by hand by applying a little water to smoothen them (as depicted in Fig. 10). Almost everyone in the so-called "tribal family" is engaged in

[Harvesting]   [Cutting & Drying]   [Bundling]

[Making Yarn manual/ hand driven & packaging]

[End products]

Fig. 10 Process flow of Sabai grass products

58

A. Khandual and S. Sahu

making ropes—such as fine (1 + 1 ply), medium (2 + 2 ply) and rough (multi-ply). The tightening and twisting of ropes are done in a traditional machine called a *Gharadi* in Oriya. Oriya is spoken in Odhisha—previously Orissa—in India. Both the state government and central government recognize the need to enhance the socioeconomic development of tribal peoples who make their livelihoods from sabai grass and, as a consequence, have set up planning commissions to look into this. Figure 10 is from Sahu et al. [37] who gave a pictorial representation of the processes and equipment used. Sabai rope is a readily available raw material used for many applications. Made of low-cost natural bio-fibres the ropes are widely preferred over synthetic threads for binding, tying and packaging (e.g., in the supply of bamboo to paper mills). Sabai rope is also being used for common household needs such as cots, tables, chairs and tapestries. Often, jute-dyed threads are used in the creation of artefacts and in decorative handicrafts.

The method of fibre extraction from sabai grass to spin yarn and weave fabric for apparel use remained unclear until recently. The method is similar to the extraction of banana fibre by means of soaking and rot steeping which is known to take place in southern parts of India, though not reported scientifically. In this process the leaves are initially crushed and beaten by a rotating wheel set with blunt knives to make sure that only the fibres remain. This is followed by retting in water and mechanical processing. Some of these techniques can be viewed on YouTube for banana [38] and coconut [39]. This has generated much commercial interest in making yarn and fibre from sabai grass as a result of its suitability for weaving, knitting and producing textiles and fabrics [40, 41]. India's Khadi and the Village Industry Commission, in its project profile for *gramodyog rojgar yojana* (Village Industry Earning Plan), estimates that from a total investment of around US$12,000 there would be net surpluses of 118, 94, 82 and 70 % on investing 100, 80, 70 and 60 %, respectively [42]. (Khadi is a term for hand-spun and hand-woven cloth.)

8.2 Processing Possibilities Such as Blending and Dyeing

As is evident from the agro-economics of sabai grass fibre, the social need for its development, its chemical constitution though FTIR and XRD, its thermal properties from DSC and DTA analyses, its properties compared with those of cotton and flax and chemical (Table 6) and morphological studies, sabai grass as a bio-fibre can be considered a good option when looked at from the sustainability perspective. Compared with coconut and jute, it can also be considered a bio-fibre of superior quality for geo-textiles and composites. Since, it is cellulosic in origin, in principle it could easily be dyed with the same kinds of dyes used for cotton.

9   Conclusion

The sabai grass industry plays a dominant role in shaping the economic status of rural people in many districts of India, China, Nepal, Indonesia and some other Asian countries. The industry involves cultivating the grass, processing goods such as ropes, mats, carpets, three-piece suites, tapestries as well as sophisticated fashionable articles. This morphological, physical, thermo-chemical compositional study has demonstrated the improved properties of this fibre once its non-cellulosic components have been extracted, removed or eliminated. With breaking strength and moisture regain properties close to cotton and flax coupled with acceptable modulus, stiffness and moderate antimicrobial properties, sabai grass fibre is a worthy alternative cellulosic fibre to other grass plants, in addition to that it is relatively easy to grow. Being a renewable, sustainable fibre, with fibre characteristics that are essential for industrial fibres, composites, geotechnical applications, etc., this golden grass fibre possesses tremendous potential for enhancing the economic status of local society. However, further research is clearly needed to clarify the social implications of making such a healthy, responsible, sustainable and techno-economic choice.

References

1. IYNF Coordinating Unit (2015) International year of natural fibres 2009. Rome, Italy. http://naturalfibres2009.org/en/iynf/index.html. Accessed 23 Oct 2015
2. Bledzki AK, Gassan J (1999) Composites reinforced with cellulose based fibres. Prog Polym Sci 24(2):221–274
3. University of Cambridge, Department of Engineering (2015). http://www-materials.eng.cam.ac.uk/mpsite/short/OCR/ropes/default.html. Accessed 23 Oct 2015
4. Xu Y, Rowell RM (2011) Biofibers, sustainable production of fuels, chemicals, and fibers from forest biomass. January 1, 2011, 323–365. doi:10.1021/bk-2011-1067.ch013
5. Bledzki AK, Faruk O, Sperber VE (2006) Macromol Mater Eng 291:449–457
6. Zou D, Chen X, Zou D (2013) Sequencing, de novo assembly, annotation and SSR and SNP detection of sabaigrass (Eulaliopsis binata) transcriptome. Genomics 102:57–62
7. Huang Y, Zou DS, Wang H (2004) The benefit of soil and water conservation of Eulaliopsis binata. Chinese J Eco-Agric 12:152–154
8. Duan WJ, Zou DS (2006) The amelioration of microenvironment of purple soil in south China by planting Eulaliopsis binata. Ecol Environ 15:124–128
9. Zhang YD, Zhang JZ, Zhang XY (1993) On prospects of the development and utilization of the resources of Eulaliopsis binata. J Wuhan Botanical Res 11:273–279
10. Liu, XH, Zou, DS, Yang, SY (2004) Effects of sun-shading at different levels on fiber quality of Eulaliopsis binata. J Hunan Agri Univ (Nat Sci) 30:107–109
11. Han P, Song GJ, Xu ST, Sun GB, Yuan F, Han YH (2008) Properties of new natural fibers: Eulaliopsis binata fibers. J Qingdao Univ (E&T) 23:44–47
12. Huang Y, Zou DS, Wang H, Yu YL, Luo JX (2003) Ecological benefit of Eulaliopsis binata grown in slope wasteland. J Agro-Environ Sci 22:217–220
13. Duan WJ, Zou DS, Luo JX (2003) The soil and water conservation efficiency of Eulaliopsis binata in the deserted sloping field of purple soil in South China. J Hunan Agric Univ (Nat Sci) 29:204–206

14. Yao JL, Zhou Y, Hu CG (2007) Apomixis in Eulaliopsis binata: characterization of reproductive mode and endosperm development. Sex Plant Reprod 20:151–158
15. Mittal SP, Sud AD (1993) Eulaliopsis binata—an important grass species of Shiwalik Hills of Northern India. J Hubei Agric Coll 13:205–208
16. Yao JL, Hong L, Zhang YD et al (2004) Genetic diversity and classification of ecotypes of Eulaliopsis binata morphological traits and AFLP markers. Sci Agric Sin 37:1699–1704
17. Liu CH, Zhou ZX, Zhou Y et al (2006) RAPD analysis of Eulaliopsis binata in different populations. Acta Bot Boreal Occident Sin 26:915–920
18. Chand N, Tiwary RK, Rohatgi PK (1988) Resource structure properties of natural cellulosic fibres—an annotated bibliography. J Mater Sci 23:381–387
19. Chand N, Sood S, Singh DK, Rohatgi PK (1987) Structural and thermal studies on sisal fibre. J Therm Anal 32:595–599
20. Tang J et al (2013) Bioresour Technol 129:548–552
21. Han JS (2015) Properties of Nonwood Fibers. http://originwww.fpl.fs.fed.us/documnts/pdf1998/han98a.pdf. Accessed 5 Nov 2015
22. Muhammad JBJ (2008) Studies on the properties of woven natural fibers reinforced unsaturated polyester composites M.S. Thesis, Universiti Sains Malaysia
23. Tyagi CH, Dutt D, Pokharel D (2004) Studies on soda and soda-AQ pulping of Eulaliopsis binata. Indian J Chem Technol 11:127–134
24. Momin SA (1950) Pakistan J. Sci. 2:42–46
25. Hadi A Biodegradable polymers. doi:http://dx.doi.org/10.5772/56230
26. Kabir MM et al (2012) Chemical treatments on plant-based natural fibre reinforced polymer composites: an overview. Compos Part B Eng 43(7):2883–2892
27. Tharanathan RN (2003) Biodegradable films and composite coatings: past, present and future. Trends Food Sci Technol 14:71–78
28. Rowell RM, Han JS, Rowell JS (2000) Characterization and factors effecting fiber properties. Nat Polym Agrofibers Compos, Sãn Carlos, Brazil
29. Chand N, Rohatgi PK (1992) Potential use, mechanical and thermal studies of sabai grass fibre. J Mater Sci Lett 11:578–580
30. Tang J, Chen K, Xu J, Li J (2011) Effects of dilute acid hydrolysis on composition and structure of cellulose in eulaliopsis binata. Bio Resour 6(2):1069–1078
31. Qu LJ et al (2005) The mechanism and technology parameters optimization of alkali-H2O2 one-bath cooking and bleaching of hemp. J Appl Polym Sci 97(6):2279–2285
32. Gierlinger N et al (2008) In situ FT-IR microscopic study on Enzymatic treatment of poplar wood cross-sections. Biol Macromol 9(8):2194–2201
33. Stewart D et al (1995) Fourier-transform infrared and Raman spectroscopic study of biochemical and chemical treatments of oak wood (Quercus rubra) and barley (Hordeum vulgare) straw. J Agr Food Chem 43(8):2219–2225
34. Tian M et al (2014) J Fiber Bioeng Inform 7(2):181–188
35. Reddy N, Yang Y (2009) Properties of natural cellulose fibers from hop stems. Carbohyd Polym 77(4):898–902
36. Qu C, Wang S (2011) Macro-micro structure, antibacterial activity, and physico-mechanical properties of the mulberry bast fibers. Fibers Polym 12(4):471–477
37. Sahu SC, Pattnaik SK, Dash SS, Dhal NK (2013) Fibre-yielding plant resources of Odisha and traditional fibre preparation knowledge—an overview. Indian J Nat Prod Resour 4(4):339–347
38. https://www.youtube.com/watch?v=nhZW2n-rqys. Accessed 5 Dec 2015
39. https://www.youtube.com/watch?v=SmnYiEDHtts. Accessed 5 Dec 2015
40. http://www.juteyarnbraid.com/jute-fabrics.html. Accessed 7 Dec 2015
41. http://trade.indiamart.com/details.mp?offer=2340138830. Accessed 7 Dec 2015
42. http://www.kvic.org.in/pmegpwebsite/pmegpwebsite/kvic-regppmegp.in/commonprojectprofile/FIBRE%20BAN%20MAKING%20UNIT%20%28SABAI%20GRASS%29.pdf. Accessed 11 Dec 2015

Potential of Ligno-cellulosic and Protein Fibres in Sustainable Fashion

Kartick K. Samanta, S. Basak and S.K. Chattopadhyay

Abstract Fashion can be encapsulated as the prevailing styles manifested by human behaviour and the latest creations by the designers of textile and clothing, footwear, body piercing, decor, etc. Fashion can trace its history to the Middle East (i.e., Persia, Turkey, India and China). Natural fibres such as silk, wool, cotton, linen, jute and ramie (a flowering plant in the nettle family) and man-made fibres such as regenerated rayon, cellulose acetate, polyester, acrylic, bamboo, and soy protein are intensively used for the production of traditional to specialty apparel, home furnishings and interior decorative textiles. To prepare fibres for use they are enhanced during spinning, weaving, knitting and chemical processing. Linen/flax is considered the most important and useful natural fibre as far as fashion is concerned for tops, shirts and summer dresses. Recently (as of 2016), a few more protein fibres —such as angora, pashmina and yak—have also been exploited to produce luxurious fashionable textiles, owing to their exotic features. Natural fibre–based textiles are being increasingly dyed in a sustainable manner using eco-friendly natural dyes that are fixed by using bio-mordants (plants that accumulate alum in their leaves). Similarly, the potential naturally coloured cotton has for traditional to fashionable end applications is also highlighted in this chapter. As far as sustainable development is concerned, textiles are preferred to be made of natural fibres and to be value-added with eco-friendly chemicals and auxiliaries, preferably derived from natural resources such as plant/herbal extracts, bio-materials, bio-polymers and bio-molecules.

K.K. Samanta (✉)
Chemical and Bio-Chemical Processing Division, ICAR-National Institute of Research on Jute and Allied Fibre Technology, Kolkata 700040, India
e-mail: karticksamanta@gmail.com

S. Basak · S.K. Chattopadhyay
Chemical and Biochemical Processing Division, ICAR-Central Institute for Research on Cotton Technology, Mumbai 400019, India

© Springer Science+Business Media Singapore 2016
S.S. Muthu and M.A. Gardetti (eds.), *Sustainable Fibres for Fashion Industry*, Environmental Footprints and Eco-design of Products and Processes, DOI 10.1007/978-981-10-0566-4_5

1 Introduction

Clothing is essential for human beings and has a long history of use (over 7000 years). Fibres such as marijuana, hemp, paul and bamboo have been reported as having been used for such a purpose. Prehistoric peoples created their own culture, which has continued to evolve and develop right up to today (2016). As far as basic raw materials of the eco-fashion industry are concerned, natural fibres are key to growth of its sustainability. Natural fibres—such as silk, wool, cotton, linen and jute—and man-made fibres—such as regenerated rayon, cellulose acetate, polyester and acrylic—are used in the production of traditional to specialty apparel, home and interior decor textiles by giving them added value during fibre spinning, weaving, knitting, non-woven production and chemical finishing. Elegant textile products are currently being used in the fashion industry to make dresses for film stars, high-class executive casual and apparel wear, world-class luxurious interiors of airports, rail and ship carriers and for furnishing five-star hotels. Similar to vegetable fibres, protein fibres like wool and silk were also used in large quantities for the production of royal apparel and home furnishing textiles for royal families. At the time of the Byzantine Empire, fabrics made of silk were considered the most valuable luxurious products, as they were the very expression of power, wealth and aristocracy. Such luxurious and fashionable fabrics were mostly used in making secular clothes, religious vestments and interior furnishings and, even today, they are very popular in the Italian market. Recently, a few additional protein fibres—such as angora, pashmina and yak—have been utilized to produce luxurious fashionable textiles owing to their exotic features such as fineness, warmth, softness, desirable aesthetic attributes, elegance, whiteness and unique hand akin to sheep wool. Up to the 1950s these natural fibres were mostly used in the production of fashionable textiles; later on, synthetic fibres slowly penetrated the market due to advances in polymer science, material science and textile science resulting in the engineering of new polymers for fibre formation with improved characteristics and performance, high durability and low interaction with the environment. However, as a result of many positive aspects of natural fibres over petrochemical-based synthetic fibres— such as bio-degradability, renewablity, eco-friendliness, higher moisture regain, good moisture absorption and desorption, soft feel, adequate to fair strength, good appearance after chemical treatment, availability in large quantity and carbon neu- trality—the demand for natural fibres is steadily increasing for textile application [1].

For the sustainable development of high-end textile products, specialty apparel and home textiles, natural fibres should preferably be used. They should then be processed and finished with eco-friendly chemicals and auxiliaries, preferably derived from natural sources. This will result in adding extra value to natural products, while preserving natural resources. In this regard, a number of bio-materials, bio-polymers, bio-molecules and bio-extracts—such as enzymes, natural dyes, aromatic and medicinal plant extracts, chitosan (a derivative of chitin), aloe vera, neem (a tree in the mahogany family), lignin, silk sericin (a protein), grape and mulberry fruit extract, banana pseudostem and peel sap, citrus oil and many more—have been extracted from agro-waste, plants and animals for the production of sustainable products for health care, skin care, wellbeing, comfort and UV-protective and flame-retardant functional textiles. The present chapter reports the same in detail [2–4]. Sustainability is the latest buzzword and covers the field of fibre, fabric, fashion, economics, food and agricultural production. The most common definition of sustainability is the "processes or products, which meet the needs of today's society, without compromising resources of the needs of future generations" [5]. Indeed, this means that for something to be sustainable it needs to be able to continue for a long time, without damaging the environment, society or becoming too expensive to continue such that one day it has to stop. Hence, for a fibre, fabric or fashion design to be sustainable, it requires to be produced without harming the environment, the people involved in the production value chain or without costing so much that one day it will not be economically viable to produce it. Sometimes it is assumed that "sustainable fibre" means an organic fibre or a natural one. It is also true that some man-made or synthetic fibres may be similar or more sustainable than natural ones as they do not use as many resources as natural fibres [5]. In the context of sustainable fashion, not only should basic raw material (i.e., the fibres) be sustainable, but the entire textile value chain, including chemical processing, dyeing, finishing and recycling/bio-degradation, is also expected to be sustainable in terms of water, energy and chemical conservation and effluent generation. Another dimension of sustainable fashion concerns the working conditions in textile and garment factories, which are often associated with long working hours, exposure to hazardous chemicals used in bleaching and dyeing processes and the scourge of child labour [1]. In the context of luxurious, fashionable textiles and specialty wear, clothing with smart attributes—comfort, soft feel, wrinkle free, light weight, pleasant, skin friendly, fragrant, contaminant free and easy on the eye—is preferred.

This chapter briefly discusses the physical, mechanical and end use characteristics of important natural fibres—such as jute, ramie, flax, hemp, banana, pineapple, wool, silk, yak, angora, and pashmina. The production of decorative/diversified fashionable yarns and fabrics from such fibres either in pure or various blended forms—such as jute/cotton, jute/banana, jute/ramie, jute/flax, jute/pineapple, jute/yak, cotton/ramie, angora/cotton, angora/silk and yak/wool fibres—has been reported in detail along with the characteristics of fashionable products—such as apparel, home textile, utility textile, lifestyle products, bags, shoes, blazers, jackets and so on—made from them. These fibres originate from natural resources and, moreover, are dyed with eco-friendly natural dyes with a bio-mordant. Similarly, the potential of naturally pigmented cotton for conventional to high end fashionable textile applications has also been reported.

2 Fibre, Fabric and Fashion

As already pointed out, "fashion" can be defined as the prevailing styles of human behaviour and the newest creations by designers of textiles, clothing, footwear, body piercing, decor, etc. Fashion originated in the Middle East (i.e., Persia,

Turkey, India and China) and a few decades later spread to Europe. Initially, fashionable products were only affordable by the royals or the rich; however, as civilization progressed, at least since the end of the 18th century, slowly such products became affordable to the middle classes and finally to the people of the world. Fabrics have been an integral part of human lives and has a long historical existence (over 7000 years); the utilization of fibres such as marijuana, hemp, paul and bamboo has been reported for such purpose. For example, in Thailand, there is evidence of animal bones and crab shells being used in the extraction of fibres and spinning at that time. Indeed, this is an early example of how prehistoric people created their own culture—one that continued to develop right up to today [6]. As civilization has developed in the last few centuries, people have become increasingly concerned about their lifestyle, fashion, health, hygiene, medicine, food and drink, comfort, luxury, leisure and wellbeing. People in the rich to super-rich category surround themselves with luxury items such as iconic cars, mansions, posh flats, expensive home furniture, jewelry, paintings, sculptures, top-quality clothing and home textiles as part of their modern lifestyles. In this context, luxurious apparel and home furnishings also play important roles and are regarded on a par with other iconic items, providing the fashionable attributes with a functional touch. Natural fibres such as silk, wool, cotton and linen and man-made fibres such as regenerated rayon, cellulose acetate and polyester are used to a great extent in the production of such specialty apparel, home furnishings, and interior decor textiles by adding extra value during spinning, weaving, knitting, and during non-woven and high-end chemical finishing [7]. Of the various plant-based cellulosic fibres, abaca (*Musa textilis*), cotton, flax/linen, bamboo and soy protein fibres dominate in the fashion industry as a result of their light weight and strength. Of the cellulose fibres, fashion designers prefer linen/flax as the most important and useful wearable natural fibre for tops, shirts and summer dresses. It should be mentioned that linen has a number of important properties: original shine, moisture absorption, allows the skin to breathe, cooling sensation and comfortable fit next to the skin. Egyptian, Scottish and Irish linen fibre–made products are very popular with fashion designers for their unique white color and outstanding shine. Similarly, jute is used for fashionable and decorative applications—mainly ornamental products—owing to its elegant natural golden colour and other fibre properties. In the fashion industry, young designers now offer 100 % carbon-neutral collections that strive for sustainability at every stage of their garments' lifecycles: from production, processing and packaging to transportation, retailing and ultimate disposal [1]. Preferred raw materials include well-known natural fibres like flax and hemp that can be grown without agrochemicals; moreover, the garments produced are found to be durable, recyclable and bio-degradable [8]. High-value textile products are currently being used in fashion and reality shows, in making dress materials for actors and actresses, casual wear for top professionals, luxurious interiors of airports, rail and ship carriers and in furnishing five-star hotels. Recently, environmental and health concerns are behind the boom in organic cotton that not only has adopted biological practices of pest control without the use of any chemical fertilizers or genetically modified seed, but also is processed without using any chemical dyes or formaldehyde. In 2007, organic cotton was grown on almost 50,000 ha of farmland in 22 countries, with total production estimated to be around 60,000 t [1]. Mainstream fashion designers and clothing companies are now slowly introducing organic cotton in the form of jeans and sportswear. Global retail sales of organic cotton clothing and home textiles were reportedly worth more than $3 billion in 2008. Much like cotton and linen textiles, protein fibres like wool and silk in ancient times were also used in large quantities for the production of clothing and home furnishings for royal families due to their unique attributes: lustre, smoothness, shine, paper-like feel, silky appearance as well as being warm and soft. As mentioned earlier, at the time of the Byzantine Empire, fabrics made of silk were considered the most valuable luxurious products, as they were the very expression of power, wealth and aristocracy. Such luxurious and fashionable fabrics were mostly used in making secular clothes, religious vestments and interior furnishings and, even today, they are very popular in the Italian market. Similar to organic cotton fibre, fashion collections often feature organic wool (from sheep that have not been exposed to pesticide dips). Similarly, cruelty-free wild silk (harvested, unlike most silk, once the moths have left their cocoons) is increasingly used. The other important protein-based natural fibres for the fashion industry are cashmere, angora, yak and their blends with cellulosic, lignocellulosic and other protein fibres. Sustainable fashion intersects with the fair trade movement, which offers producers in developing countries a higher price for their natural fibres and promotes social and environmental standards in fibre processing. As far as synthetic fibres are concerned, rayon and lyocell (a form of rayon), cellulose acetate and polyester fibres dominate the fashion industry due to their lustrous, pliable, soft, absorbent, and wrinkle-free (e.g., polyester) properties. Acrylic fibres are cheaper and satisfy many of the properties of wool fibre. Spandex, an elastomeric fibre, is utilized by many textile industries for making sportswear and exercise wear as a result of its excellent elastic/stretchable property.

3 Natural Fibre, Fashion and Sustainability

Natural fibres occupy centre stage of the current fashion movement in terms of being sustainable, green, ethical, eco-friendly and even eco-environmental [1]. Naturally coloured cotton, organic cotton, organic wool, wild silk and even flax and hemp are the important fibrous materials as far as sustainable fashion is concerned. Natural fibres put due emphasis on fashion for the environment, the wellbeing of fibre producers and consumers and the working conditions of the textile industry. The production of sustainable traditional to fashionable textiles also needs the associated raw materials, chemicals and the product value chain to be sustainable in terms of water, energy, cost, fibre and bio-degradability. Approximately 100 L of fresh water are used to process one kilogram of cotton or similar textile from preparatory processing to finishing, which is finally discharged as an effluent contaminated with residual dyes, pigments, salts, acids, alkalis, sizing ingredients, suspended solids and other auxiliaries. The discharge of such effluent into water streams has a serious consequence on flora and fauna, besides adversely affecting the fertility of agricultural land. Shortage of water in the near future will have a serious impact in the textile, agriculture, energy and allied industries. In the past, as a result of environmental norms and associated government legislation, various technologies have been developed, validated and/or implemented in the textile production/processing arena, so as to ensure reduction in water, energy, production costs and effluent load. These include (i) low material-to-liquid ratio processing; (ii) spray and foam finishing; (iii) use of enzymes; (iii) natural dyeing; (iv) digital printing; (v) infrared dyeing and drying; (vi) radio-frequency drying; (vii) ultra-sound dyeing and dispersión; (viii) dyeing with supercritical carbon dioxide; (ix) plasma processing; and (x) UV and laser-assisted processing. The chemicals and auxiliaries used in large quantities in the textile chemical processing industry are caustic soda and other alkalis, acids, salts, dyes, pigments, oxidizing agents, reducing agents, hypochlorite, sizing material, and stain removers (carbon tetra-chloride). Many of these textile chemicals and auxiliaries contain suspended solids and produce large quantities of effluent with high biological oxygen demand (BOD) and chemical oxygen demand (COD) values. Chlorinated compounds such as pentachlorobenzene, hexamethylene biguanide and quaternary ammonium are used in antimicrobial, rot resistance and moth-proof finishing. Similarly, phenyl salicylate, benzophenone and benzotriazole–based chemicals are used in UV-protective finishing. Some present day chemicals and auxiliaries used in textile processing have an adverse effect on the environment and users in terms of health and hygiene (e.g., synthetic dyes like azo are sensitive to the skin and have a carcinogenic effect). Due to increased global awareness of environmental pollution, climate change, carbon footprint, health and hygiene in the last two decades, the demand for organic material such as organic fruits, vegetable, crops and pulses, and organic cotton has been exponentially growing. In the context of sustainable fashion, natural fibres are gaining in importance owing to their advantages of bio-degradablity, renewablity and carbon neutrality as well as for possessing such properties as good moisture regain, soft feel, adequate to fair strength, and good appearance after chemical treatment [9]. Natural fibres will play a key role in the emerging "green" economy by reducing carbon emissions and recyclable materials that ultimately minimize waste generation [9]. During processing, natural fibres mainly generate organic wastes and leave residues that can be used for electricity generation or to make ecological housing material. For example, in a UN Food and Agriculture Organization (FAO) study, it was estimated that 10 % of the energy needed for the production of one tonne of synthetic fibres is required for the production of one tonne of jute fibre [1]. This is mainly because jute is cultivated by small-scale farmers who farm traditionally and the main energy input is only human labour—not fossil fuels. Jute, hemp and bamboo are often considered more sus-tainable fibres owing to the requirement for fewer or no chemical herbicides, pesticides and fertilizers. The processing of some natural fibres can lead to gen-eration of high levels of water pollutants (as reported above), but they consist mostly of bio-degradable compounds. In contrast, chemicals including heavy metals are frequently released in the effluent during synthetic fibre processing. Most natural dyes require mordant for better colour exhaustion, fixation and desired fastness. Harda (*Terminalia chebula*), natural alum and vinegar have been explored as bio-mordants for the natural coloration of textiles. For the sustainable development of traditional to high end luxury and fashionable textiles, they should preferably be made of natural fibres and then processed and finished with eco-friendly chemicals and auxiliaries, preferably derived from natural sources. This will add extra value to natural products while preserving natural resources. In this regard a number of plant extracts, bio-materials and bio-polymers—such as enzymes, natural dyes, aromatic and medicinal plants, chitosan, aloe vera, neem, lignin, silk sericin, grape and mulberry fruit extract and citrus oil—have been explored for the production of sustainable, hygienic, wellbeing-related, skincare-related, comfortable, self-cleaning, and UV-protective textiles, which are reported below in detail. Another dimension of sustainable fashion is concern for the working conditions of employees in the textile and the garment industries as they are often associated with long working hours, exposure to hazardous chemicals used in bleaching and dyeing factories and child labour [1].

4 Important Sustainable Lignocellulosic Fibres and Blended Fabrics

4.1 Sustainable Lignocellulosic Fibres

4.1.1 Jute Fibre

Jute is one of the most affordable natural lignocellulosic fibres. It is a quite strong, stiff, shiny and long vegetable bast fibre produced from plants in the genus *Corchorus* [10, 11]. The fibre is composed mainly of cellulose, hemicellulose and lignin and classified in the bast fiber category (i.e., derived from the bast or skin of the plant). Other similar fibres are kenaf (*Hibiscus cannabinus*), industrial hemp, flax (linen) and ramie. Two varieties of jute—*Corchorus olitorius* L. and *Corchorus capsularis* L.—are mainly cultivated in Asia and some parts of Africa and Latin America [12]. Around 90 % of total jute worldwide is produced in India, Bangladesh, China and Thailand. India is the major producer of this fibre—the second most important natural fibre after cotton fibre [12]. After retting, jute fibres look off-white to brown and sometimes golden due to the presence of minerals. It primarily consists of alpha-cellulose (61 %), hemicellulose (24 %) and lignin (11.5 %). The fibre has the advantages of having properties such as biodegradability, cost-effectiveness, elegant natural golden color, good strength and annual renewability. The fibre's properties are reported in Table 1. Despite having such important features the fibre is comparatively coarser than cotton, is mainly utilized for packaging of agricultural crops, food grains, and other commodities in the form

Table 1 Physical and chemical properties of some stem fibres [15–19]

Property	Jute (white)	Jute (tossa)	Kenaf	Roselle	Flax	Linseed	Ramie	Sunnhemp	True hemp	Banana
A. *Fibre physical properties*										
1. Ultimate cell										
Length (mm)	0.8–6.0	0.8–6.0	2–11	1.5–3.5	26–65	20–30	20–25	5–20	5–50	0.9–4.0
Breadth ($\times 10^{-3}$ mm)	5–25	5–25	13–34	10–32	10–35	0.3–0.5	15–80	12–36	10–40	12–33
Length/Breadth value	110	110	140	100	1700	–	3500	450	900	100
2. Filaments										
Gravimetric fineness (tex)	0.9–4	2–5	3.5–5.5	2.8–5.5	2.5–6.0	4–10	0.4–0.8	5.5–17	5–15	3–12
Tenacity (g/tex)	30–45	35–50	30–45	25–40	45–55	53.2	40–65	30–40	35–45	30–40
Breaking extension (%)	1.0–1.8	1.0–2.0	1.0–2.0	1.0–1.8	2.5–3.5	4.0	3.0–4.0	2.5–3.5	1.5–2.5	1.8–3.5
Torsional modulus ($\times 10^{10}$ dyne/cm^2)	0.25–1.3	0.25–1.3	0.65–1.4	0.6–1.2	0.8–1.1	–	0.7–1.3	1.0–2.0	0.6–1.5	0.33–1.2
Flexural modulus (dyne/cm^2)	3.0–5.5	3.5–6.0	3.5–6.5	3.5–6.5	1.8–2.5	–	0.8–1.2	125–175	25–80	20–25
Transverse swelling in water (%)	20–22	20–22	20–22	20–22	20–24	30–36	12–15	18–20	20–22	16–20
3. Bundle tenacity (g/tex)	13–30	16–35	16–30	12–30	30–36	10–35	28–40	15–35	20–35	20–30
4. True density (g/cm^3)	1.45–1.52	1.45–1.52	1.47	1.45	1.55	–	1.56	1.53	1.54	1.35
5. Apparent density (g/cm^3)	1.23	1.23	1.21	1.22	1.44	–	1.44	1.34	1.35	0.62
6. Moisture regain (%) at 65 % r. h.	12.7–13.7	12.7–13.7	13	12.5	7	8–9	6.5	10.5	11.0	15
7. Coefficient of friction; parallel & perpendicular	0.54 & 0.47	0.45 & 0.39	–	–	0.44 & 0.39	–	0.61 & 0.50	0.50 & 0.40	–	–
8. Degree of crystallinity (%)	50–50	55–65	–	–	55–70	–	66–74	45–50	–	45–50

(continued)

Table 1 (continued)

Property	Jute (white)	Jute (tossa)	Kenaf	Roselle	Flax	Linseed	Ramie	Sunnhemp	True hemp	Banana
B. Fibre chemical properties										
(i) Alpha-cellulose (%)	61.0	60.7	31–72	–	64.1	–	86.9	78.3	–	61.5
(ii) Pentosan (%)	15.9	15.6	–	–	–	–	3.6	3.6	–	14.9
(iii) Uronic anhydride (%)	–	5.9	–	–	–	–	5.6	1.7	–	5.3
(iv) Acetyl content (%)	2.9	3.5	–	–	–	–	0.6	1.5	–	2.8
(v) Lignin content (%)	13.2	12.5	8–19	–	2.0	–	0.5	4	–	9.7
(vi) Minor constituents such as fat and wax; nitrogenous matter; ash (%)	0.9; 1.56; 0.5	1.0; 1.87; 0.79	–	–	1.5; 3.9; 1.1	–	0.3; 2.1; 1.1	0.5; 1.4; 0.3	–	1.4; 1.6; 4.8
(vii) Degree of polymerization of alpha-cellulose and hemicellulose	–	1150 and 166	–	–	4700 and 120–150	–	5800 and 150	1002 and 49	–	1300
(viii) Elemental percentage (C, carbon; O, oxygen; other elements)	C 62.72; O 37.11; Si 0.06; S 0.08	–	–	–	–	–	–	–	–	–

of hessian and gunny (burlap) bags. In the last few decades with the advent of synthetic polymers—such as nylon, polyester, polypropylene and acrylic—and fibre technology, jute fibre is currently facing serious challenges for its traditional end use. However, more recently, it has found a promising application in technical textiles (e.g., in carpet backing, home textiles, geotextiles, agrotextiles, composites and auto-textiles [13]). In addition to its traditional use in packaging, jute fibre has a number of value-added end applications in making decorative and fashionable yarn and fabric, either in the pure form or in blends with other fibres—such as flax, ramie, wool, polyester and acrylic (as described in succeeding sections).

4.1.2 Ramie Fibre

Of the different fibre crops that are commercially important, ramie (*Boehmeria nivea*) occupies an important place [14]. It occupies first place among all commercial fibres because of its superior strength, as well as other desirable quality attributes such as fibre length, durability, absorbency and lustre, which make it a very useful fibre for manufacturing a wide variety of textiles and cordage products. Ramie has exceptionally long ultimate fibre cells (an average length of about 150 mm). The fibre is highly lustrous and has exceptionally high resistance to bacteria and fungi, including mildew [15]. Ramie fibre is very important in the context of moisture management, as it absorbs and releases moisture quickly, with almost no shrinkage and stretching. In undegummed ramie fibre 68.6 % cellulose, 13.1 % hemicellulose and 1.9 % pectin are present. On the other hand, in degummed ramie fibre about 96–98 % alpha-cellulose with very few trace amounts of lignin is present, when calculated on the dry weight basis. Despite several unique positive fibre attributes, it is still not considered a major textile fibre, mainly because of issues related to production and processing difficulties in large quantities. The word "ramie" comes from an ancient Malayan word. In Dutch it is *rameh*. In China it is known as *tchou-ma*, *chu-ma*, *ch'u*, *tsu*; in India it is *rhea, pooah, puya*. It is one of the oldest plant fibres cultivated in the Orient and is used in the Far East as a textile fibre of great quality. Ramie is known to have been grown in China for many centuries, even before cotton (*Gossypium* spp) was introduced by the Chinese in 1300 AD. The fibre is mentioned in ancient Indian literature like *Ramayana* and *Shakuntala*, a well-known drama written by Kalidas in about 400 AD. Yet, ramie remains a minor crop, with world production probably never exceeding 130,000 t. The major ramie-producing countries are China, Brazil and the Philippines, the others are Japan, Indonesia, Malaysia and India. All products that are manufactured from cotton, flax, hemp or silk could also be manufactured from ramie. Due to its unique textile properties, textile products made of ramie cannot be replaced by any other natural or synthetic fibre. Some of the important applications of ramie fibres are found in the making of premier-quality shirts and suits, knitwear, bed sheets, twines and threads, pulleys, belts, fire hoses, water-carrying bags, gas mantles, meat packaging, canvas, filter cloths and defence products—such as ammunition belts, camouflage nets and parachute cords. Figure 1

Fig. 1 Picture of different important lignocellulosic fibres

shows some different lignocellulosic fibres while Table 2 gives the physical and chemical properties of some seed, leaf and fruit fibres.

4.1.3 Banana Fibre

Cotton, jute, flax, ramie, hemp, sisal, wool and silk are the important fibres widely used throughout the world for the manufacture of clothing as well as home and technical textiles. Apart from these widely utilized natural fibres a large number of other fibres are also grown in different parts of the world in much smaller quantities to meet the demand of local economies. One such fibre is banana, which is extracted from the leaf sheath (pseudostem) of the banana plant—a member of the monocotyledon family [15]. Banana plants (*Musa sapientum*) are grown in different parts of India. After harvesting the bananas the trunk or trunk sheath is commonly considered agro-waste [20]. It is estimated that post-harvest a large quantity (60–80 t/ha) of agro-biomass is generated causing pollution of the local environment during disposal [17]. The fibre is obtained from the sheath of the trunk after scotching, followed by washing in water or dilute chemicals. India is the largest producer of banana fibre globally. Total global production of banana fibre is between 80,000 and 90,000 t [15]. The price of banana fibre is US$0.43–0.81 per kilogram compared with hemp, kenaf and flax, which are US$0.15–0.60, 0.15–30 and 0.15–0.21 per kilogram, respectively [17]. The fibre has many properties resembling those of jute, hence the attempts have been made to produce yarn and fabric blended with jute using the jute-spinning system to develop different diversified products (as discussed below). However, prior to mechanical processing, it needs to be stapled for effective processing in jute-spinning machineries. Due to its high cellulose content, banana fibre has tremendous scope for making superior quality paper and as a reinforcing material in natural fibre bio-composites. The cultivation of bananas for clothing and other household end uses in Japan dates back to the 13th century.

Table 2 Physical and chemical properties of some seed, leaf and fruit fibres [15–19]

Property	Sisal	Coconut	Pineapple	Aloe	Manila hemp	Cotton
1. Ultimate cell						
Length (mm)	0.5–0.6	0.5–4	3–9	0.5–5.0	3–12	15–60
Breadth ($\times 10^{-3}$ mm)	5–40	7–30	4–8	5–35	10–32	15–20
Length/Breadth value	150	95	450	125	250	1300
2. Filaments						
Gravimetric fineness (tex)	16–35	25–50	2.5–6.0	10–25	20–35	0.1–0.3
Tenacity (g/tex)	40–50	15–35	25–45	25–40	35–45	20–45
Breaking extension (%)	2.5–4.5	8–20	2.5–4	3–10	2–3	6.5–7.5
Torsional modulus ($\times 10^{10}$ dyne/cm^2)	0.3–1.0	0.2–1.5	0.3–1.0	0.2–1.5	0.3–1.2	0.8–1.2
Flexural modulus (dyne/cm^2)	125–175	150–250	2.5–4	100–150	150–200	0.3–1.0
Transverse swelling in water (%)	18–20	5–15	18–20	16–20	18–22	20–22
3. Bundle tenacity (g/tex)	22–36	10–15	20–30	15–30	20–35	–
4. True density (g/cm^3)	1.45	1.40	1.5	1.47	1.45	1.55
5. Moisture regain (%) at 65 % r. h.	11	10.5	13	12.0	9.5	7.0
6. Coefficient of friction (parallel)	–	–	0.62	–	–	–
7. Coefficient of friction (perpendicular)	–	–	0.57	–	–	–
8. Degree of crystallinity (%)	40–45	–	55–60	–	–	–
9. Chemical composition						
(i) Alpha-cellulose (%)	63.9	32–44	69.5	–	–	83–90
(ii) Pentosan (%)	7.9	–	17.8	–	–	–
(iii) Uronic anhydride (%)	5.8	–	5.3	–	–	–

(continued)

Table 2 (continued)

Property	Sisal	Coconut	Pineapple	Aloe	Manila hemp	Cotton
(iv) Acetyl content (%)	4.6	–	2.7	–	–	–
(v) Lignin content (%)	8.6	40–45	4.4	–	–	<2
(vi) Minor constituents such as fat & wax; nitrogenous matter; ash (%)	0.7; 0.8; 0.7	–	3.3; 0.25; 0.9	–	–	–
(vii) Degree of polymerization of alpha-cellulose and hemicellulose	726 and 165	–	1178 and 116	–	–	–
(viii) Elemental percentage (C, carbon; O, oxygen; and other elements)	–	–	–	–	–	C 53.2; O 46.8; S 0.08

4.1.4 Flax/Linseed Fibre

The common flax plant is a member of the small family Linaceae, which includes about a dozen genera and several species, widely distributed in the temperate and subtropical regions of the world [15]. *Linum usitatissimum* is the only member of the family used for fibre production. The seed of the flax plant is known as linseed from which linseed oil is extracted. The fibre extracted from the straw after harvesting is suitable for manufacturing coarser quality textiles. The plant is cultivated worldwide in countries as diverse as Belgium, Japan, Kenya, Uganda, Argentina, Canada and India with annual production of about 773,000 t [15]. Linen is laborious to manufacture, but the fibre is very absorbent and garments made of linen are valued for their exceptional coolness and freshness in hot weather [21]. The flax variety grown for fibre production has excellent textile properties such as good length, fineness, softness, good density, lustre and moisture absorbency. Fibre fineness (tex) varies in the range of 4–10 and fibre tenacity between 45 and 55 g/tex, whereas these values for jute are 1.2–4 tex and 30–45 g/tex, respectively. Detailed fibre properties are reported in Table 1. The fibre has slightly better elongation characteristics than jute fibre. Fine and regular, long flax fibres are spun into yarns for linen textiles. More than 70 % of linen goes to the clothing manufacture sector, where it is valued for its exceptional coolness in hot weather—the legendary linen suit is a symbol of breezy summer elegance [9]. Linen fabric maintains a strong traditional niche among the highest quality household textiles, such as bed linen, furnishing fabrics and interior decor accessories. In 2007 the European Union produced 122,000 t of flax fibre, making it the world's biggest producer followed by China with about 25,000 t. The bulk of linen production has shifted to Eastern Europe and China, but niche producers existing in Ireland, Italy and Belgium continue to supply high-quality fabrics to Europe, Japan and the USA. The fibre comprises 70 % cellulose, absorbs moisture from the air and allows the skin to breathe, with no irritating or allergenic effects. Aditya Birla Co. has been in the market place promoting flax-based clothing under the brand name Linen Club for quite some time. To provide the superior feel, look and bright colour the fabric is dyed and finished using the latest European technology. Such textile substrates are to some extent limited to the rich owing to their relatively higher price than other lignocellulosic textiles. Furthermore, flax fibre can be used for making table wear, clothing, surgical thread, sewing thread, elegant bed linen, kitchen towels, tapestries and artist canvases.

4.1.5 Pineapple Fibre

Pineapple (*Ananas cosmosus*) fibre is obtained from the leaves of the pineapple plant which belongs to the Bromeliaceae [15]. Its name is derived from the Spanish word *pina* meaning "cone shaped". The plant is widely cultivated in the Philippines, Malaya, Thailand, Ghana, Kenya, Mexico, Taiwan, China and India, with a total cultivated area of about 84,300,000 ha [15]. Brazil is the world leader in pineapple cultivation, followed by Thailand and the Philippines, supplying 52 % of global total output [22]. The leaves of the plant are about 3–5 ft long and 2–3 in. wide, tapering to a point akin to a sword. The fibre extracted from waste pineapple leaves is known as pineapple leaf fibre (PALF). The fibre is extracted from the leaves by hand as well as with a decorticator (a machine that strips skin, bark or rind of plants), though a combination of water retting and scraping is used in practice. The fibre has a fineness similar to jute fibre, but a tenacity that is significantly lower (Table 2).

4.1.6 Sunnhemp Fibre

Sunnhemp, a natural cellulosic bast fibre, is obtained from *Crotalaria juncea,* which is grown in India and neighbouring countries like China, Korea and Bangladesh as well as in Romania and Russia. It is a coarse, strong fibre that is brown-yellow in colour; the fibre is extracted after retting of the plant [15]. World production of this fibre is approximately 200,000 Mt. The fibre is fine and white in colour if it is harvested at the pre-flowering stage, but production remains 2 % lower in that case. The fibre gives a low lignin content (4 %) compared with jute, which has about 13 % lignin [15]. The fibre properties have been reported in Table 1. The fibre is coarser than jute and has comparable tenacity. Hemp has been used for centuries in making rope, canvas and paper. Long hemp fibres can be spun and woven to make crisp, linen-like fabric used in clothing, home-furnishing textiles and floor coverings [9, 21]. In China, hemp is degummed for processing on flax or cotton machinery. It is often blended with cotton, linen, silk and wool to give a soft, aesthetic feeling and to improve product durability. Pure hemp has a similar texture to linen. Hemp fashionable jewelry is the product of knotted hemp twine, done by macramé, and includes bracelets, necklaces, anklets, rings, watches and other adornments [21]. Different stitches are used to produce a wide range of fashionable hemp jewelry.

4.2 Jute-Based Fashionable/Decorative Fabrics

Jute is mainly used for packaging of agricultural crops and commodities. In the early days, the jute packaging system had a monopoly in the world market. India used to contribute the major share of foreign revenue by exporting jute goods to Western countries. However, over the years, jute fibre has faced serious challenges from advances in synthetic polymers and fibres as well as in the field of material science [12]. Due to the introduction of lightweight synthetic polymers like polypropylene and polyethylene bags in developed countries in the early 1970s, Indian traditional jute-packaging products faced stiff competition in both national and international markets. India's Central and State Governments, the jute industry and research organizations put much emphasis on developing alternative products to packaging bags and discover other applications for jute and jute-blended (with natural and synthetic fibres) textiles. The use of synthetic polymeric fibres currently used in large quantities in place of natural fibres is now raising concern about environmental pollution, the preservation of natural resources and biodiversity. Hence, natural fibre–based textile products are once again in market demand. The main advantages of natural fibres—such as jute, cotton, flax, ramie, wool, silk and banana—are their biodegradability, renewability, cost-effectiveness, natural golden colour (as well as other colours), good strength, good moisture regain, very good thermal insulation, and non-toxicity. Over the last few decades, research and product development have been intensified to produce different jute-blended ornamental textiles and to establish a complete value chain from fibre to fabric or fibre to fashion. The jute industry is having a demand for diversified products in value-added end applications such as floor covering, upholstery, handicrafts and fashionable items [23, 24]. The production of decorative jute fabric with ornamental fashion attributes deserves special mention as far as handloom-made jute products are concerned. In this context, Roy and Basu reported the development of suitable and durable jute yarn, fabric and specialty textile-based products (bags/packs) for alternative or value-added end applications [25]. Such specially designed, engineered products were named "spun-wrapped yarn" or "covered yarn" when used in the hand-weaving machine. As expected, jute spun-wrapped yarns showed much lower hairiness, better tensile and flexural performance than those made from traditional jute yarn [25]. The same research group also described the development of

Fig. 2 Full view of the handloom developed with a Jacquard-shedding arrangement for diverse product and ornamental fabric development [24]

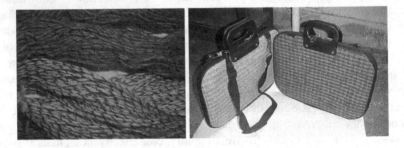

Fig. 3 Jute yarns and fabric products [12]

a handloom with a Jacquard-shedding arrangement with necessary modifications for the development of fashionable jute-blended fabrics (Fig. 2) [24]. They reported the development of various jute yarns and fabric products, such as a laptop carry bag with a price tag of INR 300 ($US4.44) and a school bag of INR 250 ($US3.67) from specialty jute yarn (Fig. 3) [12].

4.2.1 Blended Yarns

The blending of different fibres has primarily been adopted to improve the technological performance of major components and/or to improve the economic properties of yarn produced from such blends.

4.2.2 Fine Yarns

Fine yarns (84–207 tex) are needed to produce finer and stronger fabrics for furnishing, upholstery and industrial applications. Long, fine natural fibres—such as ramie, flax and PALF—are commonly blended with jute for such purposes.

An array of specialty yarns from jute and blends with other fibre/filaments/films has been developed at National Institute of Research on Jute and Allied Fibre Technology (NIRJAFT) in Kolkata (India). A simple gadget was designed, developed and fabricated for use in the existing spinning frame for manufacturing jute-covered yarn. Dyed viscose multifilaments were wrapped around the jute during spinning to mask the jute core fully. Moreover, such yarns were used to weave decorative fabrics with smooth and colourful surfaces. Polypropylene monofilament–covered yarns were also used to weave high-performance fabrics suitable for industrial use. High-density polyethylene (HDPE) slit film–covered jute yarns were spun to produce water-resistant fabric. Core spinning technology was also adopted to spin core yarns with HDPE/HDP core and jute sheath. These yarns had higher dry and wet strength and elasticity as well as improved evenness. All these properties were reflected in fabrics woven from jute/synthetic core yarns as well. Novelty/Fashionable yarns were produced from coloured synthetic tops and dyed jute slivers to produce fabrics for various end uses. Different filament-covered jute or jute-blended decorative/fancy yarns (276 tex) were also developed for a similar purpose. Computer bags, ski bags, school bags, office bags, upholstery, lifestyle products and garments were manufactured from various jute and/or blended yarns and fabrics as shown in Figs. 4 and 5.

Jute fibre is a major cash crop in India and is mainly used for the production of coarser packaging fabrics for the packing of rice, wheat, sugar, potatoes, onions, etc. However, ordinary jute yarn cannot be used for the development of decorative and upholstery fabrics. It is possible to develop high-value utility products—such as school bags, laptop bags, office bags, ladies' bags and folder files—from jute by

Fig. 4 Fashionable jewelry products made of jute fibre and fabrics

Fig. 5 Fashionable lifestyle products developed from jute-blended textiles

using specialty jute yarns. Fabrics developed with a different weave structure to manufacture such products, compared with ordinary jute fabric of similar constructional parameters, showed higher strength and elongation, but a drastic reduction in the bending modulus, resulting in a softer feel. Similar new products can also be developed from other natural fibres akin to jute—such as *Hibiscus cannabinus* (kenaf), *Hibiscus sabdariffa* (rossel), *Cannabis sativa* (hemp) and *Linum usitatissimum* (flax).

4.3 Jute-Blended Yarns and Fabrics for Fashionable/Diverse Applications

4.3.1 Jute–Ramie Fibre Blends

Degummed ramie contains approximately 4.3 % gum, whereas decorticated ramie contains 23 % gum. Blending of ramie with other natural or synthetic fibres can eliminate its inherent drawbacks, making it suitable for fabric formation for casual as well as formal wear, tablecloths, handkerchiefs, canvas, suit cloth and mat edging. It has been reported that 10–15 % blending of raw ramie (with 25–30 % gum content) or partially degummed ramie (with 9 % residual gum) with jute helps to spin good-quality yarns of finer count (100–105 tex) in jute or flax machinery [18]. Blending of ramie with jute helped to produce yarns of 103 tex, which is not achievable with wholly jute fibre. As the gum of ramie is gradually removed, spinnability and yarn characteristics were found to improve accordingly. To process ramie on jute-spinning machinery, it was found that a degumming treatment with residual gum content of 8 % might be sufficient. Indeed, as there is no specialized spinning system available in India, it was attempted to spin ramie fibre on the jute system [26]. The properties of jute/ramie–blended yarns of were found to be much better than those produced from completely jute fibres. Such blended yarns have potential application in such areas as furnishings, upholstery and clothing fabrics. Wholly ramie yarn spun in the jute-spinning system also holds a lot of promise for use in shoe canvas, soles, sewing twines, coarse fabrics, etc. Binary blending of degummed ramie has been optimized with fine fibres such as viscose, polyester, silk and tussar (a type of silk) waste. In another approach, degummed ramie fibres were cut into 40-mm staple lengths with a staple cutter. Flock blending of ramie and cotton fibres in different proportions has also been carried out. Ramie is highly appreciated for its lustre and strength, whereas cotton is well known for its fineness and elongation properties. Therefore, cotton–ramie blends have been spun using the short-staple spinning technique. A blend ratio of 65:35 of cotton:ramie has been found to give an adequate count strength product (CSP) for 40 s ring yarns intended to be used for making towels and knitted products. Ramie has been blended with polypropylene and acrylic fibres in various proportions at different stages of processing to obtain suitable blend proportions for specific end uses and for ascertaining the right stage of blending. Fabrics with different area densities have been woven from various combinations of cotton and ramie yarns blended with acrylic or jute for making safari suits, shirts and other clothing. T-shirts have also been designed and tailored from fabrics woven from ramie/acrylic–blended yarn and cotton yarns.

4.3.2 Jute–Flax Fibre Blends

Flax fibre from Belgium was blended with jute fibre to produce a good-quality yarn to be used as shoe twine and clothing-grade textiles. The blend was processed in a rove spinning system with a wet spinning attachment fixed on the spinning frame. Highly regular, strong and fine yarns (84–138 tex) could be manufactured with a jute:flax blend ratio of 50:50 [18]. Subsequently, flax tow (coarse, broken fibre) was blended with jute using a small-scale jute-spinning system developed by NIRJAFT, and yarns of 138–207 tex were spun with 50 % flax tow in the blend with jute. These yarns hold much promise to be used as furnishing fabrics. Flax/jute–blended yarn of 138 tex was also prepared successfully on a conventional jute-spinning system.

4.3.3 Jute–Cotton Fibre Blends

In the processing of jute–cotton fibre blends, jute fibre was first required to shorten and then blend with cotton [18]. Jute was initially cut and mixed with cotton fibre in the stack-blending technique. Then the blend was processed in the khadi-spinning system using such machines as a bale opener lap maker, comb-bladed carding, fixed flat metallic carding, drawing, apron drafting and the *Ambar Charkha* (sky wheel). (Khadi is a term for handspun and hand-woven cloth.) Then fabric was produced on a handloom. When jute was blended with cotton fibre in the cotton-spinning system the meshes of jute reed had to be broken first in the jute-carding machine to ensure a staple length of 25 mm. The jute fibre was subsequently blended, carded, drawn and spun in the cotton-spinning system. It was possible to mix 20 % of jute fibre successfully with cotton in the cotton-spinning system and, thus, the blended yarn could be used to weave scrim cloth. Research into the rotor spinning of jute/cotton–blended yarn was also carried out by Doraiswami and Chellamani (1993) [27]. Additionally, efforts have been made to standardize the spinning parameters for open-end spinning of a 50:50 blend of jute–cotton (with 31-mm staple length) to make a yarn of 312 tex at a rotor speed of 50,000 rpm and delivery speed of 80 m/min [18]. A comparative study was conducted to evaluate the physical properties of a jute/cotton–blended curtain compared with a 100 % cotton curtain [28]. Three different blend ratios—jute:cotton at 60:40, 50:50 and 40:60—were prepared; it was found that the strength of the blended curtain warpwise was close to the 100 % cotton curtain before washing, but the strength decreased after washing. Wrap-spun jute yarns with linear densities of 276, 190 and 120 tex with wrap density in the range of 250–450 wraps/m were produced with a 2-ply cotton yarn as the wrapping element using hollow spindle technology [29]. Instead of cotton as a wrapping component, wrap-spun jute yarn of 276 tex was also developed using a viscose rayon multifilament as the wrapping element.

4.3.4 Jute–PALF Fibre Blends

Special techniques have been adopted for processing PALF (which comes from the pineapple *Ananas comosus*) on the jute-spinning system. Similar to jute processing, PALF was first softened with a 15 % mineral-oil-in-water emulsion applied at 1 % of the weight of the fibre for improved spinning performance [18]. A jute finisher card and a full circular flax finisher card with progressively higher pin density were used for first and second carding, respectively. Blending of PALF with jute was found to improve the quality of the blended yarns noticeably. A minimum 10–15 % of PALF in the blend was enough to produce a finer yarn which cannot be produced by jute fibres alone. Blending of jute–PALF was also undertaken at NIRJAFT to investigate the extent to which the superiority of PALF may be utilized for upgrading the performance of jute yarn. It was found to produce stronger and finer yarn than yarn based just on jute fibre making it effective at producing diverse textiles. A suitable technique was developed at the institute to process the fibre in jute machinery. Finisher card slivers of jute and PALF were mixed in different proportions at the first drawing stage, keeping variables such as machine sequence and processing parameters identical. The performance of jute/PALF–blended yarn was improved in terms of strength by increasing the proportion of PALF fibre in the yarn. PALF has been used as raw material for manufacturing a number of products such as paper, rope, handkerchiefs, knitted shirts, interlining lace, mats, bags, blankets, insulators, soundproofing material, nanomaterial and composites [30].

4.3.5 Jute–Banana Fibre Blends

Banana fibre is obtained from the sheath of banana trunk after scotching and washing in water or in a diluted chemical. The fibre has many attributes similar to those of jute fibre, hence the attempts to produce yarns and fabrics adopting the jute-spinning system. However, the fibre needs to be stapled first for efficient processing in jute-spinning machine sequences. Due to its high cellulose content, banana fibre has tremendous potential for use in the production of good-quality paper and as a reinforcing material in the preparation of green-composite, fine-quality fancy yarn and decorative fabrics [20]. The possibility of blending Indian varieties of banana (*Musa sapientum*) sheath fibre with jute using the jute-processing system was explored by Sinha (1974) [31, 32]. White jute, tossa jute and kenaf were blended separately with 75 and 50 % banana sheath fibre at the jute finisher carding stage [18]. Yarns of 345 and 280 tex were spun, where quality was found to deteriorate marginally by increasing the proportion of banana fibre in the blends. The yarn can be used as hessian weft and sacking warp. Subsequently, an attempt was made to produce rope from banana fibre. This comprised 40 % banana fibre, 50 % aloe fibre and 10 % sisal waste tow [31, 32]. The performance of all these yarns was compared with a normal commercial agricultural rope containing about 40 % kenaf (mesta) fibre in place of banana fibre. Normal rope-making machinery, which consisted of a jute softener, teaser card, sisal tow breaker card, sisal tow finisher card, first-passage screw grill drawing, second-passage screw grill drawing, gill spinner apron draft (AD) for rope yarn, roll winder, rope-stranding machine and rope-laying machine were used. Before processing, the fibres were treated with an oil–water emulsion. Primary yarn of 24 s

Table 3 Tensile properties of jute/banana fibre–blended yarn [20]

Type of yarn	Yarn linear density—actual (nominal) (tex)	Twist (tpi)	Tenacity (cN/tex)	Breaking extension (%)	Work of rupture (mJ/tex-m)
100 % jute	–	–	10.9	2.0	0.96
Jute/banana (75/25)	268 (276)	4.0	9.0	1.4	0.68
Jute/banana (50/50)	272(276)	4.0	8.3	1.5	0.57
Jute/banana (25/75)	270 (276)	4.0	7.4	1.3	0.52
100 % banana	352 (345)	3.5	7.4	2.4	0.88

Fig. 6 Jute/banana fibre–blended fabric and clothing textiles (jackets) [20]

rope count (4600 tex) made of banana fibre instead of mesta was found to be stronger and more extensible than standard commercial rope made of aloe (50 %), kenaf (40 %) and sisal waste tow (10 %).

At NIRJAFT, extensive research has been conducted into different aspects of banana fibres such as fibre quality evaluation, processability and product development. Trials involving the spinning of yarn (6 lb, 5 tpi, in apron drafting) for blends of 100 % banana and jute/banana fibres (75/25, 50/50 and 25/75) have also been carried out. The spinning performance of 100 % banana and jute/banana (25/75) blends were not encouraging; however, spinning trials for production of coarser yarn (8 lb, 4 tpi and 10 lb, 3.4 tpi on a slip draft spinning frame) from jute/banana fibres in the same blend ratios and 100 % banana fibre yarn (10 lb, 3.4 tpi using a slip draft spinning frame) were successful. It was postulated that 100 % banana fibre can be processed smoothly in jute-processing machines and can be spun to 10-lb grist (345 tex) and above count [20]. The tensile properties of 100 % jute and banana, and jute/banana blended yarns are reported in Table 3. Different jute/banana fibre blended products are shown in Fig. 6.

4.3.6 Jute–Wool Fibre Blends

Exhaustive research work has been undertaken into spinning chemically softened jute, better known as "woollenized jute", in which jute fibre is treated with 18 % NaOH solution at 25 °C for about 30 min using various spinning systems [18]. It was observed that removal of crimps in woollenized jute fibre brought about a deterioration in the regularity of woollenized jute/wool–blended yarns at the carding and drawing stages when spun in the jute- or flax-spinning system. Hence, the spinning of woollenized jute/wool–blended yarn in jute- or flax-spinning machinery was not considered worhwhile. However, it is worth mentioning that good-quality yarn could be spun from a woollenized jute:wool 50:50 blend using woollen- or worsted-spinning systems. The yarn so produced is suitable for blankets, scarves, pullovers, wraps, and the face yarn of loop pile tufted carpets. Other researchers have also reported on blending wool fibre with jute using the jute-spinning system [33–35].

4.3.7 Jute–High Bulk Acrylic Fibre Blends

A simple and economical process of bulking jute/acrylic fibre–blended yarn has been developed in the laboratory. Such yarn has high potential for use as a substitute for woollen or 100 % acrylic-bulked yarn. The yarn has such properties as high bulk, high extension, low flexural rigidity and good strength compared with its parent blended yarn or pure jute yarn. After successful binary blending, the effect of blending, plying and bulking on tenacity, elongation, evenness, shrinkage, specific volume, and specific flexural rigidity of the developed yarn from ternary blends has also been studied. After several trials, a suitable blend ratio of 50:30:20 of jute: shrinkable acrylic:non-shrinkable acrylic was optimized. With this blend ratio, synthetic fibres—polyester, polypropylene and viscose—other than non-shrinkable acrylic fibre were also found suitable. Jute/shrinkable acrylic/viscose showed a shrinkage of 28 % with a specific volume of 11.8 cm^3/g. A substantial drop in tenacity and a large increase in extension on the bulking the ternary-blended yarns were also found, as expected. However, the tenacity and extension values were found to be comparable with similar-grade woollen yarns. The specific flexural rigidity of ternary-blended parent yarns was much less than pure jute yarns. Jute blended with hollow polyester fibre–bulked yarn (80:20) was prepared by chemical treatment. Warm fabrics like shawls and jackets were prepared from such bulk yarns. Nanofinishing was incorporated in the fabric to reduce bending rigidity. Nanopolysiloxane-based finishing was found to exhibit encouraging results. Jute fibre can also be blended with more than one fibre to impart the positive properties of other blended fibres to jute-blended yarn. Keeping this concept in mind, jute: shrinkable acrylic:hollow polyester (50:30:20) was successfully blended to get a bulk yarn of 275 tex for making upholstery, kurta (upper garments for male and female), shirts, ladies' waistcoats and gents' jackets.

4.3.8 Jute–Polypropylene Fibre Blends

Sengupta and Debnath described a new approach for making jute/polypropylene (PP)–blended yarn (70:30) by blending the fibres on a finisher draw frame [23]. The fabric was developed in a modified handloom at much lower cost. The fabric showed higher areal density, thickness and other mechanical properties. Soft, bulky, resilient and highly extensible jute/PP–blended yarns have been developed by texturizing yarns of different compositions and twist levels. Chemical texturization of different blended yarns has been standardized at NIRJAFT. The use of coloured PP fibre of 4–6 D (denier) in the blended yarn helped to produce a coloured texturized yarn, due to the emergence of coloured PP fibre on the surface of the yarn during texturization. Texturized jute/PP–blended yarn was found to have high extensibility, moderate strength, higher diameter and smoother surface feel compared with the parent blended yarn or the pure jute fibre yarn with similar construction parameters. A partially covered structured yarn has also been developed with jute and colored PP. It was aimed at covering jute yarn with synthetic staple fibres so as to reduce the harshness of jute yarn and improve its aesthetic appeal for use in clothing and furnishing fabrics. Two types of jute-based blended yarns were produced from jute:PP:hollow polyester (50:25:25) fibres and jute:shrinkable acrylic:hollow polyester (50:30:20) fibres in the conventional jute-spinning system [13]. The first blend could be used in making cushion cloth, mattress cloth, table cloths and bed sheets, whereas the second blend could be used for producing warm garments. Cross-laid, needle-punched non-woven fabrics were also prepared from 100 % jute and its blend with PP as a minor constituent for traditional to high-end applications [36].

4.3.9   Jute–Viscose Fibre Blends

As stated earlier, jute has limited end applications as a result of being a much coarser fibre. However, when it is blended with viscose—a very fine cellulosic apparel-grade fibre—the physical and mechanical properties of the blended yarn were found to improve, thus making it suitable for wider and diverse end applications [37]. Blended yarns of 50 % jute, 50 % viscose and 100 % polyester were used to make a yarn with three different ratios of jute–viscose and polyester (70/30, 50/50 and 30/70). A quaternary-blended plain woven fabric was produced on the handloom with cotton in the warp and jute–viscose–polyester yarn with three different ratios in the weft. The 30/70 jute viscose/polyester union fabric exhibited better performance than those produced with other ratios of the blend. It also reduced the cost of the product [37]. The technology of spinning covered yarn in the existing jute-spinning frame is quite novel, yet simple. A low-cost gadget has been developed that can be fitted to the spinning frame to produce a jute/viscose–covered yarn. A coloured viscose mul-tifilament of 150 D was wrapped on the surface such that the jute remains at the core of the yarn. Thus, a coloured jute/viscose–covered yarn was produced with a low proportion of viscose (10–30 %). This yarn was as fine as 84–130 tex with a smooth surface morphology and reduced hairiness compared with pure jute yarn of similar construction. Bamboo viscose—new kind of regenerated fibre—was also blended with jute. The fibre is known to have various small pores on the fibre surface that help improve moisture absorption and oxygen vapor permeation in garments. Moreover, the fibre exhibits excellent protection from ultraviolet radiation, infrared radiation and microbes [6].

5 Important Properties of Different Protein Fibres and Fabrics

5.1 Properties of Important Protein Fibres

5.1.1 Wool Fibre

Wool fibre is obtained from the follicles of sheep, goats, camels, rabbits and camelids like vicuna, llama and alpaca as wool fleece and is an important animal hair fibre [38]. Wool is composed of 18 different amino acids; the important ones are cysteine (13.1 %), glutamate (11.1 %) and serine (10.8 %). Grease–wool fleece contains impurities such as wool grease and perspiration products (e.g., suint) as well as adhered materials such as dirt and vegetable matter. It has three distinct morphological parts: the outer is the cuticle layer, the middle is a group of spindle-shaped cortex cells and the inner is the medulla. Wool fibre is mainly used for men's and women's woven outerwear, knitwear, underwear, socks, hand-knitted yarn, blankets, upholstery, filled bedding, rugs and carpets [38]. India exports nearly INR60 billion ($US884,819,400) worth of woollen carpets and other handicraft items [39]. Wool fibre is inherently hydrophobic in nature due to the presence of covalently bonded lipid on the surface of the cuticle membrane [40]. This can be overcome in practice by a chemical treatment with strong alkali or chlorine or plasma treatment so as to remove parts of the bound fatty acids. The treatment will also help in formation of polar functional groups on the fibre surface.

5.1.2 Silk Fibre

Silk is derived from the silkworm *Bombyx mori* and is composed of two major proteins: sericin and fibroin. Fibroin is a fibrous protein, present as a delicate twin-thread linked by disulphide bonds, enveloped by successive sticky layers of sericin that help in the formation of a cocoon [41]. Silk sericin or glue material is a globular protein that comprises 25–30 % silk proteins [42, 43]. Silk protein consists of 18 amino acids, many of which have strong polar side chains like hydroxyl, carboxyl and amino groups. Its high hydrophilicity with a moisture regain value of 11 % is due to the high content of serine and aspartic acid (approximately 33.4 and 16.7 %, respectively). Silk fibre made of fibroin has many end uses—such as textile fibres and in medical, industrial and cosmetic applications—because of its unique properties such as durability, water absorbency, dye affinity, thermo-tolerance, lustre, softness, smoothness and insulation. India is the second largest producer of raw silk after China and the largest consumer of silk-based fancy products [44]. It is an important raw material for producing precious fabrics, parachutes, tyre-lining materials, artificial blood vessels and surgical sutures [45]. The manufacture of lustrous silk from the dried cocoons of silkworm involves separating fibroin from sericin by a degumming process; the sericin is mostly discarded in the wastewater.

Sericin, which until recently was considered a waste product of the silk-processing industry, is now an important industrial material for food, pharma, cosmetics and textile end applications because of such properties as excellent moisture absorption and release, UV resistance, cell protection and wound healing; it is also used in anticancer, anticoagulant and antioxidant activities and in the inhibitory action of tyrosinase [46, 47].

5.1.3 Pashmina/Cashmere Fibre

Animal fibres (hairs) have special attributes such as fineness, softness and lustre that are rarely associated with any other vegetable cellulosic or lignocellulosic fibres. In addition to providing softness and lustre to the product, exotic hair fibres in many cases carry an elegance value. Based on the fineness and physical attributes of the hairs obtained from the goat family, they have been classified as mohair/angora, cashmere/pashmina, cashgora (crossbred) and guard/goat hair [48]. Pashmina, popularly known as "cashmere", is well known for its fineness, warmth, softness, desirable aesthetic attributes, elegance, fashion, whiteness and unique hand compared with sheep wool [49, 50]. It is a more luxurious, softer and warmer fibre than superfine merino wool. The word "pashmina" comes from *pashm* meaning "soft-gold" in the local language and "wool" in Persian. It is composed of 18 amino acids with alpha-keratin arranged in a helical structure akin to wool [50]. The amino acid composition is very similar to wool except for the presence of cystine, tyrosine (12 % more than wool) and proline (9 % less than wool). The fibre has the ability to add warmth to the fabric without addition any weight. Of the various protein fibres, pashmina is the finest animal fibre; it is produced in fairly large quantities. Pashmina consists of down fibres from the undercoat hair of the domesticated goat *Capra hircus*, which is indigenous to Asia. It is always mixed with coarser outercoat fibre better known as "guard hair". Worldwide production of pashmina fibre is about 10,000–15,000 t/yr. The major producing countries are China (70 % share), Mongolia (20 % share), Iran, Afghanistan, Pakistan, Nepal and India. India produces about 40–50 t/yr, which is less than 1 % of total cashmere production. Indian cashmere fibre is 12–13 μm in diameter, 55–60 mm in length and available in white, gray and brown colours [49]. The proportion of undercoat to guard hair is 40–50 %.

The quality of cashmere fibre produced in different countries differs significantly. The fibre is properly known by its origin—for example, Mongolian, Chyangara (Nepal) and Australian cashmere [10, 50]. The fabric is usually hand woven and made from hand-spun yarn. A specially designed ladies' woven shawl with unique embroidery work can cost as much as INR10,000–50,000 (US$147–737). Moreover, there is demand in international markets. The fibre is used to produce different fashionable textiles with functional attributes such as knitwear, scarves, shawls, blankets, gloves, hats, woven fabrics and overcoats (Fig. 7). The hair obtained from Indian Changthangi and Chegu breeds of goat is distinctly called "Pashmina fibre" or "Indian Cashmere", due to its fineness (10–14 μm), softness

Fig. 7 Value-added
pashmina shawl [48]

and warmth [48]. It is one of the costliest fibres and mostly used for the production of high-end fashion garments.

5.1.4 Angora Fibre

Angora rabbit fibre is considered one of the world's finest luxury fibres owing to such properties as extreme warmth, excellent whiteness, very good lustre, soft silky touch and lightness. Most products made of angora fibre are very expensive, reflecting the laborious harvesting process and the small number of producers [51, 52]. It occupies third place in animal fibre produced in the world after wool and mohair [52]. It is no secret that the world's softest garment fibre comes from the Angora rabbit [53]. Quiet and calm by nature, these animals have been used for fibre harvesting for hundreds of years and are thought to have originated in Turkey. The fibre is 10–20 µm in diameter and 40–70 mm in length and has air-occluded cavities that ensure high thermal insulation. Longitudinal and cross-sectional scanning electron micrographs of Angora rabbit hair show that it is somewhat scaly but has a smoother surface structure with a prominent medulla [54]. Similarly, the lower density of 1.15–1.18 g/cm^3 makes it a better choice than wool (1.33 g/cm^3) and cotton (1.54 g/cm^3) fibres for making lightweight clothing and fashionable luxury garments [53]. Angora garments are very lightweight, extremely warm and soft, hence their use in trimming sweaters and the knitting of hats and scarves.

5.1.5 Yak Fibre

Yak is one of the world's most exotic specialty rare animal fibres. The total yak population is about 14.5 million [55]. Yak herds are found in mountainous regions of Afghanistan, Bhutan, Mongolia, Russia, China, India, Kyrgyzstan, Tajikistan and Nepal on the Central Asian plateau. Mongolia has the second largest yak population in the world followed by China [54, 56]. Yak skeleton hair and down fibre are seasonal in nature; abdominal and tail hair fibres are gradually shed and replaced by new ones, ensuring the thermal balance of the body during cold seasons. Fine yak hair is very similar in appearance and fineness to cashmere and has good crimp and tensile strength (9.18 cN/tex) [55]. There are four different colours of yaks, hence yak fibre can be black (68.5 %), brown (16.9 %), blue (8.9 %) or white (5.7 %) [56, 57]. Yak hair, a rare resource of specialty animal fibre, is mainly produced in China. Fibre yield (hair) is about 410,000 t/yr; 10,000 t of which is fine hair. Yak fleece also contains a large amount of coarse hair/fibres that are quite thick and stiff [58]. Each yak produces about 100 g of down fibre annually and the fibre comes in a few natural colours, of which white is the most valued [56]. The quantity of yak fibre produced depends on factors such as the sex, age and breed of the yak [59]. The coat of the yak is composed of three types of fibre varying significantly in appearance and characteristics [59]. The proportion of different layers varies throughout the seasons. Coarse-grade fibre (79–90 μm) forms an outercoat of long hair, mostly used by nomads in tent making—which characterizes the appearance of the yak. Down fibre—the finest fibre (16–20 μm) generally shed by the animal during late spring/early summer—is suitable for textile application. Middle-grade fibre (20–50 μm) is naturally strong (but not as strong as the outer layer) and mostly utilized in making ropes and tents. Consequently, decreasing the diameter of the fibre would be a good way to improve the economic value of yak hair. Stretching slenderization is a means of modifying animal fibres which involves chemical treatment and physical drawing of fibres. Yak hair fibre has been utilized by nomads in the Transhimalaya for over a thousand years to make clothing, tents, ropes and blankets. Recently, the fibre has also been used in the garment industry to produce premium-priced clothing and accessories for well-known companies/brands like Louis Vuitton and British heritage brands like Dunhill, Eileen Fisher and Vince. Since the mid-20th century, material science experiments into yak fibre have been carried out whetting the appetite of the garment industry for yak wool, by proving its exotic nature and favourable performance attributes, making it an attractive alternative to cashmere fibre.

5.2 Protein Fibre–Blended Fashionable Yarns and Fabrics

5.2.1 Cashmere–Polyvinyl Alcohol Fibre Blends

The limited availability of specialty cashmere fibre has resulted in most of it being utilized locally. Yarns are made with the help of specially designed manual spinning wheels, locally known as *charkha* or *yander* [50]. The fibre available in India is 12–13 μm in diameter and 55–60 mm in length and comes in white, grey and brown colours [49]. Harvested fibres are traditionally spun by manual spinning wheels producing yarn of R 25/2 tex, suitable for making lightweight shawls. The proportion of undercoat to guard hair is 40–50 %. It is very difficult to spin cashmere fibre mechanically (rather than manually) owing to problems associated with softness, shortness of fibre length and slipperiness which creates lapping during the carding and spinning processes, in addition to generation of high static charges [60]. Several attempts have been made to spin cashmere wool mechanically after incorporation of another fibre, known as a "carrier fibre". One such method is blending either with nylon or water-soluble polyvinyl alcohol (PVA) fibres for spinning on a worsted-spinning system, followed by weaving. The carrier fibre is then removed to manufacture high-end fashionable textile products. The carrier nylon or PVA fibre is removed from the fabric by treating with hydrochloric acid or hot water, respectively. The PVA-based process was considered to be more eco-friendly than the nylon-based process; however, the whiteness and hand properties of such fabrics were inferior to the latter process [49]. To improve the problem of whiteness and hand an alternative method was developed, in which dilute sulphuric acid was used in place of hot water to remove PVA component fibre. Thus, it was possible to enhance the whiteness index by 28 % and the hand index by 20 % compared with the hot water-based process. Traditionally, pashmina was also manually wound on a small flange bobbin known as a *parota*. Sizing of the yarn was carried out in hank form using *saresh* (a collagen and fat-based glue) as an adhesive to improve weavability; the weaving of pashmina yarn shawls was carried out in a special type of handloom [50].

5.2.2 Angora Fibre Blends

Despite the many positive attributes of angora fibre, its full potential has yet to be reached. The reason for this is the outer surface of the fibre is very slippery, making it challenging for yarn spinning. Hence, it is most often blended with other cellulosic or protein fibres such as wool, mohair, cotton or silk for spinning, followed by production of fashionable garments, winter clothes and underwear. Besides the production of cotton/polyester blends, Indian fine short Angora rabbit hair (11.1 µm, 32.3 mm) has been blended with cotton (30:70) for production of yarn with low-shrink properties [61]. Softness and an elastomeric feel is brought about by microderm softening [62]. Such fabrics are primarily used for making such items as sweaters, mittens, baby clothes, shawls and millinery (Fig. 8) [51]. In this regard,

Fig. 8 Sweater made from Angora rabbit fibre [53]

Chattopadhyay et al. reported the blending of Angora rabbit fibre with cotton fibers. The softer feel and low-shrink properties of cotton/angora fibre–blended knitted fabrics were found to be suitable for women's underwear and children's wear [52, 61]. Some commercial blends of angora knitting yarns are:

- 70 % angora, 30 % nylon
- 50 % angora, 25 % merino wool, 25 % polyester
- 40 % angora, 50 % wool, 10 % nylon
- 70 % angora, 30 % silk
- 50 % viscose, 25 % nylon, 15 % angora, 10 % wool

To produce 100 % Angora rabbit yarn the fibre surface needs to be modified so as to introduce crimps or roughness on the surface. With this in mind, the National Institute of Design (NID), Ahmedabad (India) in collaboration with Institute for Plasma Research (IPR), Gujarat (India) developed a prototype atmospheric pressure plasma treatment for angora fibres. The treatment between 1 and 10 min could enhance surface friction. Plasma—an ionized gas composed of ions, electrons, protons and UV light—were used to increase the coefficient of friction from 0.10 in the untreated sample to 0.30 in the 1-min air plasma–treated sample [63–65]. This modification enables the production of 100 % angora fibre yarn that can be used to produce stoles, shawls, scarves, caps, sweaters and multilayer products for soldiers. Attempts have also been made by other research groups to spin rabbit hair in blends with sheep wool, viscose and polyester fibres using woollen-spinning or cotton khadi–spinning systems [52]. However, very little work has been carried out into spinning this hair fibre in blends with cotton using the short-staple cotton-spinning system. Chattopadhyay et al. explored the possibilities of producing cotton/rabbit hair–blended yarns by adapting commercial cotton-spinning systems to run at economical production speeds. Hair was blended with cotton having a 2.5 % span length of 33 mm and a micronaire value of 2.9 µg/in. [52]. Samples with angora:cotton blend ratios of 10:90, 20:80, 30:70, 40:60 and 50:50 were prepared. Increasing the angora–hair fibre content in the blended yarn led to a decrease in lea CSP, breaking tenacity and elongation of both single and double yarns [52].

5.2.3 Wool Fibre Blends

Pre-treatment of wool fibre can modify its physiochemical and mechanical properties to meet process requirements [38]. Oxidizing or reducing agents are frequently applied to wool to make it reactive, so as to ensure effective and uniform post-treatment results. The effect of protease/lipase enzyme pre-treatment, followed by polysiloxane-based combination finishing to enhance the hand properties of wool–cotton union fabric has been studied by Ammayappan and Moses [40]. It was observed that both enzymes could improve the hand of the union fabric, which could be further improved by different polysiloxane formulations.

5.2.4 Yak Fibre Blends

Coarse-grade yak (guard) fibre does not contribute much to the production of value-added products and little has been reported in the literature. This may be due to the stiffness of guard fibres and/or the slippery surface of yak hair, posing challenges to the production of yarn, either from 100 % yak fibres or as a major component in blended yarn. Moreover, yak fibre—either as fine/down or coarse/guard hair—has not been blended with jute fibre with the objective of developing jute-blended yarn and fashionable fabrics using the jute-spinning system. With this in mind, NIRJAFT, after several initial attempts, developed an 8-lb 50:50 jute (unbleached):yak guard fibre–blended yarn that can be converted into a plain woven fabric (Fig. 9). Yak is one of the few exotic fibers that tempts shoppers who have grown weary of ubiquitous cashmere [66]. Recently, the British luxury brand Alfred Dunhill introduced a small collection made of yak wool blended in equal parts with merino. Other well-known brands that have developed yak wool collections include Eileen Fisher and Vince [59]. As a result of natural high strength and coarseness the guard hairs are typically carded and then worsted spun [67]. Multiple plies of guard hairs can then be braided into ropes, halters and belts or weaved into very durable rugs and bags. In contrast, down fibre with a diameter of 14–16 µm is very soft and comparable with cashmere or camel fibre; it is processed into sliver and roving and then spun into yarns for the exotic fibre market.

Tibetan cloud worsted is a precious, natural yarn spun from Tibetan yak down fibre. It is a plied, uniform, very soft yarn that is manufactured in heather colours [68]. It is suitable for all kinds of knitwear and accessories, including men's and babies' clothing. Different premier clothing textiles made of yak down fibre blended with other natural fibres are also available: typical blends are 70 % mulberry silk/30 % yak down hair, 50 % mulberry silk/50 % yak down hair, 70 % Tibetan yak/30 % baby camel and 75 % Tibetan yak/25 % bamboo viscose.

Fig. 9 Jute/Yak fibre–blended woven textile

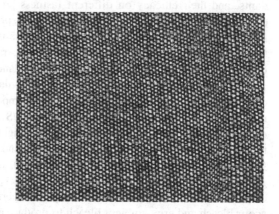

6 Sustainable Colouration of Textiles

6.1 Sustainable Dyeing and Printing Using Natural Dyes

6.1.1 Lignocellulosic Textiles

Natural dyes have been used for colouring food products, leather goods, protein fibres like wool and silk and cellulosic and lignocellulosic fibres since prehistoric times [69]. With the advent of widely available and cost-effective synthetic dyes with moderate to excellent colour fastness properties since 1856, natural dyes were slowly replaced with synthetic colours. Synthetic dyes presently are mainly used for colouration of jute and other textile substrates in the production of value-added textiles and other diverse products due to such advantages as wide colour range, different categories of dyes, mordant-free processes and availability in large quantities [70]. Dyeing such textiles with synthetic dyestuff is sometimes characterized as having a highly negative impact on the environment and its users. The use of natural dyes for the colouration of textiles worldwide has mainly been confined to craftsmen, small-scale/cottage-level dyers and printers, small-scale export houses, and producers dealing with high-valued eco-friendly textiles and sales [69]. In the recent context of health and safety, eco-concern, sustainability, carbon footprints and global warming, natural dyes and dyed textiles are once again in market demand by eco-concerned users. Therefore, the use of natural dyed textiles is steadily increasing owing to being environment friendly, having effects that are relatively less toxic and less prone to allergens, availability of a large plant base, and additional functionalities like UV protection and antimicrobial activity [60, 70]. In the case of jute fabric, some attempts have been made to elucidate the fundamental aspect of natural dyes as well as to enhance its washing and light fastness properties using several metallic mordants [69, 71]. However, little work has been reported into the application of combinations of bio-mordants and chemical mordants, and their efficacy on different fastness properties, evenness of dyeing and colour yield. Chattopadhyay et al. reported the application of natural dyes extracted from four common natural dye sources: manjistha (*Rubia cordifolia*), annatto (colouring from the seeds of *Bixa orellana*), ratanjot (*Onosma echioides*) and babool (*Acacia nilotica*). Samples were mordanted by single/double-mordanting processes using bio-mordants and chemical mordants. Bio-mordanting followed by chemical mordanting resulted in substantially improved uniformity and evenness of naturally coloured dyed jute fabric, higher K/S (colour strength) value, good to excellent wash fastness and moderate to good light fastness [70]. Pan et al. reported the application of natural colour in bleached and ferrous sulphate–mordanted jute fabric using deodara leaves (*Cedrus deodara*), jackfruit leaves (*Artocarpus integrifolia*) and eucalyptus (*Eucalyptus globulus* L.) [72]. Even deeper shaded samples exhibited good washing fastness. Grey jute fabric was bleached by grey bleach, scour bleach and grey ambient bleach to produce a white fabric prior to dyeing with direct and reactive dyes. White and dyed jute fabrics were padded with a finishing formulation comprising resin, softener and non-ionic surfactant, followed by drying and curing [73]. They found no alteration in the colour of bleached and dyed jute fabrics, before and after such value-added finishing.

Similar to the dyeing of jute fabric with natural dyes the printing of bleached jute fabric with natural dyes has also been explored after extracting the dyes using an aqueous extraction method. Scoured and bleached jute fabrics were padded with potash alum/ferrous sulphate and dried [74]. The fabric was then printed with a print paste consisting of thickener, urea and natural dye, followed by drying and steaming at 110 °C for 3 min. The printing of single- or double-mordanted jute fabric using manjistha and annatto dyes was found to be encouraging. Natural dyes when used in powder or paste form produced a better result. Wet- and dry-rubbing fastness values were found to be excellent in the case of printed jute fabric with manjistha and good in the case of annatto.

6.1.2　Cellulosic Textiles

Dyeing textiles using natural dyes is an ancient craft in India; however, it received a setback from growing research in Europe into textile coloration using synthetic dyes. Presently, a large number of synthetic dyes are used for colouration of cotton textiles, a few of which are carcinogenic in nature, as they are derived from petrochemical products [75]. Various synthetic dyes—such as reactive, direct, disperse, acid, basic, vat, sulphur, napthol and metal complex—have been developed and the efficacy of such dyes on natural and man-made textiles in terms of shade depth, number of shades, colour consistency, economy and wash durability has been thoroughly studied. Moreover, various natural dyes imparting such colours as yellow, red, blue and black are available in the market and used to produce fashionable textiles. Yellow-coloured natural dyes are extracted from turmeric, carotenoid and annatto—used to impart yellow shades to cotton textiles. The orange flowers of tesu (*Butea monosperma*), onion (*Allium cepa*) skins and marigold (*Tagetus patula*) have also been explored to see how effective they are at imparting yellow colours to cellulosic textiles. In all these cases, it was noted that mordanting was essential [76]. During the dyeing of pre-mordanted cotton with fabric marigold flower and pomegranate (*Punica granatum*) peel the dyed fabric showed adequate antimicrobial properties, in addition to imparting on attractive yellow colour [77]. The natural red colour imparted to cellulosic textiles could be produced without any mordanting of the fabric. In this regard, carthamin from safflower (*Carthamus tinctorius*) petals and bark extracts produced a strong red colour with good fastness properties. Root extract of ratanjot, Indian madder (*Rubia cordifolia*) and European madder (*Rubia tinctorium*) also imparted a similar attractive red color due to the presence of carthamin [78]. Plant extracts that contain a high amount of tannin generally produce brown to black shades. Similarly, catechu (*Acacia catechu*) imparts a dark brown colour to cotton fabric. It has been reported that—apart from tannic acid—tannin also contains catechin and quercetin that act as an astringent and antioxidant. Similarly, Wannajum and Srihanam reported on natural dyeing of bamboo fibres with indigo [6]. Fibres dyed with indigo revealed a good level of light and sweat fastness.

6.1.3 Protein Textiles

Besides its utilization for traditional textile products, wool is also used for the production of various fashionable garments. Wool and silk—being protein fibres— contain amine and carboxylic acid groups. Unlike silk, aqueous solutions of wool contain no net charge due to the presence of equal numbers of positively and negatively charged molecules. Presently, a large number of synthetic dyes are used for colouration of woollen textiles; however, in recent years the application of natural dyes to protein fibres has increasingly met the demand for green fashion. Therefore, the selection of mordants, dyeing time, temperature and pH during the colouration process using natural dyes play critical roles. Tannin-rich materials— such as harda and gallnuts—and metal salts—such as alum, aluminium sulphate and ferrous sulphate—are suitable for mordanting protein fibres. After mordanting at 80–90 °C with 5 % mordant for 30 min the fabric can be dyed under acidic conditions at 60 °C for 30 min, with a material:liquor ratio of 1:25 [79]. Mogkholratasitt et al. (2011) reported on the dyeing of protein wool fibre using leaf extract of eucalyptus, which contains tannin and quercetin as chromophoric colour materials [80]. Various kinds of natural mordant combinations—such as lemon juice and stannous chloride, lemon juice and ferrous sulphate, lemon juice and copper sulphate—have also been attempted for the natural dyeing of woollen textiles. Dyed wool fabric mordanted with those compounds exhibited better light and wash fastness. Pashmina yarns/fabrics have also been dyed using natural dyes obtained from varying sources to produce different colours: for example, annota seed for red, indigo for blue and henna and myrobolan (*Terminalia chebula*) for yellow and brown [50]. Color from babool bark has been extracted and its efficacy for woollen textiles studied [75]. Colourant extraction was found to be higher in slightly alkaline media than aqueous, acidic and alcoholic media. Similarly, alkaline extracted samples showed higher K/S values with very good fastness to light and washing than those of other media. Dried extract in different media imparted different shades of brown and grey to woollen yarn, both in mordanted and non-mordanted samples.

6.2 Camouflaged Fashionable Textiles

Active camouflage entails blending into the visual surroundings present in nature as done by several groups of animals including cephalopod mollusks, fish and reptiles. Two principal factors are responsible for the active camouflage effect in the animal body. The first is counter illumination and the second colour change [81]. Efforts have been made in the past to impart such camouflage effects to textile substrates to produce fashionable textiles. Camouflage is frequently used for defence applications—such as military dress, vehicles and helicopter painting—for safety and security purposes. It is worth mentioning that CSIRO (Commonwealth Scientific and Industrial Research Organisation) scientists have developed camouflage military fabrics by mixing different dyes that reflect visible, UV and infrared light in such a way as to mimic the reflectance spectrum of backgrounds such as plants, soils and buildings. It works best when reflected light from a fabric matches background reflection [81]. Zhang et al. (2007) studied four different vat dyes to measure the camouflage effect they had on the fabric. Application of 1–2 % C.I. Vat Blue 13 dye plays an important role in green camouflage as fabric reflectance matches the reflection of green leaves [82]. (C.I. stands for Colour Index^{TM}.) The visual effect of camouflage could be changed by altering cotton fabric construction. Recently, Goudarzi et al. (2014) reported camouflage effects on cotton fabric in the visible and near infrared (NIR) region using three different vat dyes—C.I. Vat Blue 6, C.I. Vat Yellow 2 and C.I. Vat Red 13—to get a reflectance curve similar to the green of forest leaves and NATO green shades [83]. Tulip Ltd provides a tie dye T-shirt in camouflage colours to provide a stylish fashionable look. The flat work surface of the fabric is covered and then a suitable mixture of kit colours are applied until it is saturated, followed by color fixation on the fabric for 6–8 h while it is still moist [84]. It is also possible to make a person invisible by finishing the garment with highly reflective material. The first research into this area was initiated in the 1960s by Harvard University and the University of Utah in the USA [85].

6.3 Natural Pigmented Cotton

Naturally pigmented or coloured cotton and the fine fabrics made from them for nearly five millennia in Peru probably constitute the oldest record of yarn spinning and weaving in human history [86–92]. Naturally coloured cotton is believed to have originated in America around 5000 years ago when plants were selected for different natural colours (red, green, brown and tan) other than the normal yellowish off-white. The vast array of naturally coloured cotton has been well documented since the time of the New World explorers. There were originally shades of cotton ranging from brown, dark green, black, red and blue. The natural color of cotton fibre comes from natural pigments such as caffeic acid and cinnamic acid that are deposited in alternating layers with cellulose on the outside. These varieties of cotton were widely used by Native American people. However, coloured cotton later became obscure and unavailable to markets, as farmers and manufacturers found them difficult to spin mechanically owing to their shorter staple length. The colour attributes of cotton fibre solely depend on the plant genotype and on manipulating plant genes. In India, it has been possible to produce red, green and brown shades in recent years [86–92]. The green attribute, less common than brown, occurs in two shades: light green and green. Green is not only more prone to fading than brown—it fades faster too. Prolonged exposure to sunlight during boll opening leads to rapid fading of green and finally leads to a white, off-white or brownish colour. On the other hand, the portion of lint not directly exposed to sunlight retains its original lint colour. Green is mostly observed in *Gossypium hirsutum*. To date, naturally pigmented cotton is not considered an industrial crop in India and little research in this area has been undertaken possibly due to lower farm yields. Growers are usually paid a higher price because of the lower yield of such fibre. In 1993 the coloured cotton price was in the range US$3.60–4.50 per pound compared with conventional white cotton in the range US$0.60–0.90 per pound.

Nevertheless, coloured cotton plants have higher resistance to pests and do not need any toxic pesticides. Thus, coloured cotton is also known as "eco-friendly cotton", hence it can be used for the sustainable fashion industry. Moreover, by virtue of possessing intrinsic colour the fibre does not require post colouration such as dyeing or printing. Therefore, the colouration process could also be considered sustainable by meeting the requirements of green fashion and environmental pollution norms. Despite its several advantages, coloured cotton is still not well accepted in the traditional market as a result of challenges in spinning finer count yarn from low fibre quality as well as the low availability of desired shades. Unlike conventional cotton, natural coloured cotton does not fade during laundering. On the contrary, its colour is reported to improve and become stronger. Naturally pigmented cotton, especially green cotton, has excellent sun protection properties with higher ultraviolet protection factor (UPF) values than conventional bleached or unbleached cotton. Compared with white cotton, it has been observed that brown cotton exhibits excellent antibacterial properties, with a bacterial reduction rate of 89.1 and 96.7 % against well-known gram-positive and gram-negative bacteria (Staphylococcus aureus and Klebsiella pneumoniae), respectively.

As already mentioned, naturally coloured cottons, such as brown- and green-coloured cottons, have been grown and used by humankind for some 5000 years. Coloured lint samples recovered from coastal areas of South America have been dated to 2500 BC. In India, brown, khaki and red cottons were commercially grown in specific locations like Rayalseema in Andhra Pradesh and exported up to the 1950s. Even the world-famous Dacca muslin was woven using white and coloured cotton lints. These coloured cottons suffered the drawbacks of low yield and poor fibre quality. Hence, their cultivation was abandoned, and they were eventually replaced by white cottons of higher yield and superior fibre quality. When Europe and the USA started demanding cotton textiles free from harmful dyes and pesticide residues, there was a revival of interest in organic cottons and naturally coloured cottons. Subsequently, commercial cultivation of coloured cottons was started in the USA. Indeed, owing to this renewed interest, agricultural scientists in India began trials for new coloured cotton in 1995 as part of the All India Coordinated Cotton Improvement Project (AICCIP) [87]. Assessments of such cottons for spinning potential as well as yarn and fabric characteristics are necessary to improve fibre quality. The trials highlighted the spinnability of a number of coloured cottons developed indigenously [87]. Moreover, trials looking into full-scale spinning, weaving and knitting as well as into yarn and fabric characteristics have also been undertaken. Coloured cotton was found easy to open, clean and draw due to having a higher wax content on the fibre surface, but produced inferior-quality yarns than white cotton of equivalent length, quality, fibre maturity and higher wax content. Thus, the spinnability of such cotton was limited to 24 s Ne (24 tex) maximum as a result of having a shorter fibre length (24–25.8 mm), smaller uniformity ratio (44–48) and lower micronaire value (2.7–6.1). It was found that naturally coloured cottons are best utilized in preparing fabrics, along with white cotton, to produce attractive stripe and check effects.

Despite having many environmental advantages, there are limitations to coloured cotton. It is simply not viable owing to low yield, poor fibre quality, limited colour range and limited market demand [89]. The Central Institute for Cotton Research (CICR) in Nagpur (India) and several other state agriculture universities began in 1990 looking into finding ways of improving fibre length, strength, maturity and yield in 1990. Similarly, the Northern India Textile Research Association (NITRA) began cultivating camel brown and olive green naturally coloured cottons of *Gossypium hirsutum* in its fields along with white cotton (J-34 variety). This led to various fibre and textile properties being evaluated and compared [90]. It was found that fibre length and strength of coloured cotton were inferior to J-34 (white cotton). Fibres were subsequently converted to 1/3 twill fabric of white ring-spun yarn (8 s) as warp and 8 s blended weft yarn (white and colour blended). Post-scouring, it was found that untreated samples of coloured cotton fabrics showed greater colour depth than treated samples. This may be attributed to leaching of the natural colour pigment on washing. There was a drastic change in colour on treatment with hydrogen peroxide in both the green and brown cotton samples. Post-treatment, all the brown-shaded fabrics turned white with a whiteness index of 80–84, comparable with the index (86) of white cotton. The same value for green cotton was 74–84. No difference in colour fastness to light was found both pre- and post-treatment in the case of white cotton. On the other hand, coloured cottons had a moderate light fastness rating (2–3) which could be improved to 4–5 after applying various chemical formulations. However, wash fastness in terms of staining on white fabric of the same was fairly good (3–4). As above, after treatment with suitable chemicals it was possible to improve colour fastness to washing in terms of change in shade, whereas staining on white cloth did not show significant change. Significant research has been carried out in this area at Central Institute for Research on Cotton Technology (CIRCOT) in Mumbai (India) where various fibre parameters, yarn properties and textile performance were evaluated in terms of washing and light fastness. It was found that coloured cotton contained more wax than white cotton. Other fibre properties like moisture regain, scouring loss and degree of polymerization remained similar to those of normal cotton. It was observed that scouring generally reduced colour intensity by 20–40 %, while soap washing led to an improvement in colour intensity of 25–50 %. The micronaire value of green- and brown-coloured cotton was in the range 3.4–4.1 and tenacity 32.1–36.0 g/tex for those qualities of the fibres. The advantages of natural coloured cotton are:

- Being naturally coloured, no additional dyeing is required and the process is considered eco-friendly and sustainable.
- Some varieties are inherently resistant to insects and disease and thus require fewer pesticides during cultivation.
- It has drought and salt tolerance properties.
- Fabric made of naturally coloured cotton exhibits excellent UV-protective and Sun-protective properties.
- Fibres made of coloured cotton exhibit good antimicrobial properties.
- Growers are normally paid higher prices for cultivation.

The disadvantages of natural coloured cotton are:

- Fibre spinning is challenging owing to shorter fibre lengths, hence it is normally blended with white cotton. Such hybrid yarns are used by brands like Levi Strauss, L.L. Bean, Eileen Fisher and Fieldcrest for making clothes (e.g., khakis).
- It needs to be grown at isolated farms so as to avoid any contamination with white cotton.
- It has poor light and washing fastness.

7 Textiles with Fragrance-Imparting Finishes

Globalization has resulted in people being increasingly busy in their personal and professional lives. Subjected to a raft of time-bounded responsibilities, they frequently get stressed physically and mentally. To keep stress manageable and live a healthy and relaxing life, they normally resort to exercise, yoga, leisure and trips to the spa; sleep of course is important too. Luxurious clothing and home textiles can play important roles by imparting a fresh, healthy and hygienic feel to the wearer and user of such textiles. As a result of repeated everyday use, home textiles, clothing, T-shirts, socks, bed linen, pillow covers and bed sheets do not remain fresh. Moreover, textiles constantly exposed to sweaty, hot and humid conditions offer favourable conditions for microbial growth leading to bad odours. A solution would be to get plant extracts/plant molecules/bio-molecules or synthetic chemicals that impart fragrance by incorporating them in textile products. This could provide users with pleasant and fresh surroundings by masking bad smells. Aroma is mostly composed of oils extracted from plant products and synthetic materials consisting of large aromatic molecules. Market demand for high-value, aroma-imparting, fashionable textile products is being met by Bombay Dyeing & Manufacturing Co. Ltd, which has launched a new collection of aroma-based textiles imparting a pleasant and refreshing fragrance to bed linen and decor in living rooms and bedrooms, providing much-needed relief to otherwise busy lives. They also have a collection of luxurious bed linen made of high thread count, soft cotton satin smelling of natural lavender, jasmine and rosewater which revitalizes the senses and rejuvenates the soul. Furthermore, perfumed bed linen helps to relieve mental stress, while ensuring comfortable sleep. Such finishes are applied to textiles by microencapsulation which facilitates the slow release of fragrance for lengthy periods of time. The actual use of such textiles results in the breaking of micro-capsules owing to pressure or friction and the diffusion of fragrance molecules into the air, making the surroundings fresh and pleasant [93]. Similarly, Scottish researchers have developed a microencapsulated aroma-therapeutic luxury textile that is highly beneficial for cancer patients; it involves airtight hard-shell capsules containing a particular type of aroma which have been shown to alleviate the side effects of chemotherapy and radiotherapy [94]. Functional textiles can be obtained using newly engineered fibres or by incorporating functional agents in conventional fabrics. Microencapsulation is effective at protecting these functional agents from reactions with moisture, light and oxygen. If a fabric is treated with microencap-sulated functional agents, such as an aromatic essential oil, the durability of such finishing is expected to be higher [95]. Specos et al. reported the development and post analysis of two types of microcapsules containing essential oils for application to cotton fabrics [95]. A lemon-perfumed essential oil encapsulated in GAM increased fragrance durability in cotton fabrics compared with non-encapsulated essential oil that could only withstand one washing cycle. Electronic nose analysis is an objective and adequate method for monitoring the fragrance released from microcapsules.

Jasmine, lavender and sandalwood are the preferred aromas for textiles and home furnishings; they contain active ingredients like santalols, fusanol, santene, teresantol, benzyl acetate, linalool, linalyl acetate, benzyl benzoate and geraniol [96, 97]. Besides their mind-blowing fragrance, these materials help in revitalizing the immune and central nervous system, skin nourishing, smoothening of facial lines and wrinkles, cell regeneration and are used as an antidepressant and antiseptic agent. The absorption of such molecules from different herbs into the bloodstream improves blood circulation and inhalation by clearing the throat and lungs. Clevertex has developed bath towels finished with seaweed extract and ZnO, which have been directly incorporated into the inner structure of fibres. Towels finished in this way provide comfort and relief from stress and their functional attributes are unaffected by repeated washing and wearing. Additionally, such finishes possess good antimicrobial and liquid absorbency properties resulting in a product that is comfortable and odour-free [98]. An Andrew Morgan's collection of USA recently introduced a novel aromatherapy-infused eco-friendly textile, in which tiny poly-meric microcapsule shells impart a long-lasting unique fragrance when diffused in the air [99]. In similar vein, Prince Kataria Textiles has launched a high thread count, luxurious, soft-cotton, satin bed sheet that is lightly fragranced with the pleasant aroma of natural products. This kind of aroma-embedded fabric ensures relaxation and comfortable sleep [100].

8 Sustainable Healthcare and Skincare Textiles

8.1 Healthcare Textiles

Antimicrobial health and hygiene attributes have been incorporated in textile sub-strates to control bacteria, fungi, mold, mildew and algae, as well as their associated side effects such as product deterioration, staining, odours and related health issues [101]. Hospital textiles—such as theatre drapes, gowns, masks, sheets and pillow covers—are known to be major sources of cross-infection, hence the textiles used in their manufacture need to be dressed with antimicrobial treatments prior to their use in hospital, so as to prevent or minimize the transmission of infection diseases. The term "antimicrobial treatment" refers to a broad range of technologies or applications that can provide varying degrees of protection against microorganisms. Such microorganisms have a negative economic effect on producers, retailers and users of textiles [101]. Providing customized textiles economically to address issues related to human health is the natural thing to do; however, it is a key challenge facing the textile industry that needs to be addressed appropriately. As described above, various cellulosic, lignocellulosic, protein and man-made fibres are used in large quantities in the production of sustainable fashionable textiles. Natural fibres/fabrics—such as cotton, ramie, flax, jute, silk and wool—under hot and humid conditions provide a very favourable microclimate for rapid microbial growth which results in skin diseases, foul smells and reduces the service life of valuable products. Such damage can be addressed by incorporating antimicrobial, antibacterial, antifungal, rot-resistant and moth-resistant chemicals into textile substrates. Since some of these agents cause environmental pollution because of their synthetic origin, in the last two to three decades academic and industrial research has been intensified to find alternative sustainable formulations that can provide the necessary antimicrobial efficacy at a lower cost.

Many natural materials and/or bio-materials—such as neem, nanolignin, silk sericin, aloe vera, chitosan and tulasi (*Ocimum sanctum*)—have been applied for antimicrobial, UV-protective, antioxidant, skin-nourishing, and hydrophilic finishing of textile substrates. As these materials are produced from renewable sources, they are cost-effective, easy to apply, safer to humankind and the environment. Plants containing phenols and oxygen-based derivatives are considered secondary metabolites that can act as antimicrobial and insecticidal agents [96, 97]. Similarly, tannins (naturally occurring polyphenols) are also responsible for the antibacterial activity of natural dyes. The antimicrobial efficacy of chitosan, aloe vera, neem, tea oil, eucalyptus oil and tulasi leaves on various natural and synthetic textiles has also been investigated. Chitosan is derived from chitin and is an effective natural antimicrobial agent due to the presence of an amine group that can easily react with the negatively charged bacterial cell wall and finally destroy it. Similarly, chitosan citrate has been explored for durable-press and antimicrobial finishing of cotton textiles. Thilagavathi et al. (2010) identified the active antimicrobial molecules in neem, pomegranate and prickly chaff flower that can control the growth of microbes

[102]. Neem leaves contain limonoid-based azadirachtin, salannin and nimbin that are actually responsible for antimicrobial and insecticidal properties in textile substrates. The combined effect of neem and chitosan in the form of neem–chitosan nanocomposites has also been utilized and applied in cotton textiles for eco-friendly antimicrobial finishing [103]. A recent patent on microencapsulation of neem oil and its application to cellulosic and blended textiles showed good antimicrobial efficacy. Joshi et al. (2007) reported the antimicrobial properties of neem seed extract on polyester/cotton–blended textiles using glyoxal, aluminium sulphate and tartaric acid in the two-dip, two-nip method on a padding machine; the treated fabric exhibited excellent antimicrobial activity against both gram-positive and gram-negative bacteria for five washing cycles [104]. Recently, Ahamed et al. (2012) developed a new method of preparing antimicrobial textiles (i.e., by herbal coating in neem extract nanoparticles) [105]. Nano herbal extract–treated textiles were found to show excellent antimicrobial activity against both gram-positive and gram-negative bacteria and the finish was durable up to 20 washing cycles, in contrast to 10 washing cycles for the neem extract–treated sample. Tulasi has been well known for its medicinal effects since ancient times and for its capacity to cure or resist many infections and diseases. Tulasi leaf extract contains caryophyllene, phytol and germacrene antimicrobial compounds; its efficacy on cotton textiles after methanol extraction has been studied. Tulasi-dyed bed sheet fabric shows good antimicrobial properties; it has been used as a bedding material for patients suffering from chest colds, coughs, itchiness and mucus problems. Sathianarayanan et al. (2010) applied tulasi leaf and pomegranate extract to cotton textiles using a number of methods: direct application, cross-linking and microencapsulation [106]. Methanol extract and pomegranate molecules showed 99.9 % reduction in bacterial growth on cotton fabric when applied directly by the pad dry method. As expected, microencapsulation and cross-linking of such formulations led to better results in terms of wash durability (15 washing cycles) of the finish. Recently, aloe vera gel—another important industrial bio-material—has been utilized on cotton textile to improve antibacterial efficacy against *Staphylococcus aureus* [107, 108]. Specimens treated with 5-gpl aloe vera gel showed excellent antimicrobial activity in terms of greatly reduced number of colonies and clear zone of bacteria inhibition. The antimicrobial finish imparted was durable to 50 washing cycles (slightly decreased to 98 %). Henna and juglone (obtained from black walnut) contain napthaquinone that acts as an antibacterial and antifungal agent. Curcumin has been used as a natural dye and an antibacterial agent for woollen textiles. Similarly, cotton textiles treated with turmeric, cumin, clove oil, karanja (*Millettia pinnata*), cashew shell oil and onion skin have shown good antibacterial properties. Such fabrics can also be used for medical purposes and casual clothing. Gupta et al. (2006) investigated the antimicrobial activity of cotton fabric treated with tannin-rich extract of *Quercus infectoria* (QI) in combination with different mordants—such as alum, copper and ferrous sulphate [109]. QI extract (12 %) on its own showed antimicrobial activity of 40–60 % against both the gram-positive and gram-negative bacteria. After application of the same plant extract to cotton textiles with 5 % alum and 1 % copper sulphate, antimicrobial activity was found to be significantly enhanced (70–

90 %). Samples treated with QI on its own lose antimicrobial activity after five washing cycles, whereas many more washing cycles could be sustained when treated with a mordant.

Ammayappan and Moses reported the antimicrobial efficacy of aloe vera, chitosan and curcumin either alone and in combination with each other in cotton, wool and rabbit hair fibres using the exhaustion method [51, 110]. Aloe vera showed better antimicrobial efficacy than chitosan and curcumin when applied on its own. It was possible to enhance its efficacy by adding both chitosan and curcumin. Application of the three antimicrobial agents together to peroxide-treated cotton and formic acid–treated wool/rabbit hair substrate was found to be durable up to 25 washing cycles. In similar vein, Raja and Thilagavathi evaluated the effect of enzymes and mordants on antimicrobial finishing of woollen textiles [111]. Akin to the mordanting process of wool fabric to improve colour fastness and dye uptake, it was found that enzyme treatment could also improve dye uptake. The antimicrobial activity of natural dyed textiles could be significantly influenced by the presence of mordant and enzyme treatments. Jute is susceptible to microbial attack under normal atmospheric conditions owing to the presence of hemicelluloses, hence it is quickly degraded by microbes; a detailed degradation study on this was performed by Basu and Ghosh [112]. Humidity, warmth, and a medium pH of 6.5–8.5 are favourable conditions for microorganisms to degrade jute, much like the case with cotton. Finishing chemicals preventing the growth of various microorganisms on textile substrates are the bactericides which destroy bacteria and fungi and bacterial state/fungal state inhibitors that inhibit their growth. The most important antimicrobial agents are phenolic compounds, quaternary ammonium salts and organometallic compounds [10].

8.2 Skincare Textiles

In recent years, wipes rather than conventional woven fabrics have increasingly been used in skincare, cosmetics, surface-cleaning and other skin applications due to their being less expensive, more hygienic, softer, disposable and more convenient to use [113]. Other benefits are smaller size, lighter weight, high flexibility and ease to use as well as being easy to carry and allowing quick action and drying [97]. Wipes are used to clean the body; treat wounds, rashes or burns; and to provide a fresh energetic feel to the skin. Wipes can be made of paper, tissue paper or non-woven fabric; subjected to mild rubbing or friction they remove dirt, oil and liquid and/or release soothing fragrances, medicinal ingredients and moisturizing agents. Wipes made of tissue paper or non-woven fabric are subsequently soaked with different skincare/nourishing products as discussed below. Superwipes made of Recron® bi-component yarn have been introduced by Reliance Industries Ltd; they are specially designed to hold a lot of water and provide improved cleaning capacity [114]. Johnson & Johnson has also marketed baby wipes specially designed to take care of the tender skins of newly born babies. It is much softer than other wipes and contains baby lotion to provide adequate moisture to the skin [115]. Aditya Birla Group has also marketed wipes under the brand name "Kara", suitable for face care, hand care, skin care and baby care [116]. These products are engineered with various natural ingredients—such as jojoba, avocado, honey, almonds, aloe vera, cucumber, mint and chamomile—for their outstanding effectiveness at preventing skin infections as well as soothing and moisturizing as a result of the presence of essential nutrients. They soften the skin; reduce wrinkles and fine lines (signs of ageing); strengthen skin tissue; keep the skin hydrated, nourished and refreshed by effectively removing dirt and excess oil; and provide antiseptic and anti-inflammatory properties. As these products are made of cellulose and finished with natural ingredients, they are completely natural, biodegradable and sustainable. Aloe vera (*Aloe barbadensis* Miller) belongs to the Liliaceae family and has been utilized for cosmetic and medical applications. It has excellent skincare properties that include anti-inflammatory and anti-ageing attributes. Kimberley-Clark Inc. Ltd. has patented an aloe vera application as an anti-ageing and moisturizing agent. In similar vein, DyStar Auxiliaries GmbH has developed a textile product containing a mixture of vitamin E, aloe vera and jojoba oil in a silicon matrix for moisturizing and UV-protective finishing of different textile substrates [46, 117]. Silk protein (sericin) holds a lot of potential for biomedical applications because it imparts oxygen permeability to textiles, protects cells from UV radiation and microbes and has antioxidant, moisture regulation and wound-healing properties [97, 118]. Lenzing recently launched Tencel® C fibre, which is basically a chitosan-soaked tencel fibre. Chitosan is the second most available natural polymer after cellulose and has a long history of use in cosmetics and pharmaceuticals for relieving (from itching), regulating and protecting skin and for antibacterial finishing. Stockings made with Tencel C have been reported as protecting the skin, allowing it to retain more moisture, improving the skin and stimulating skin cell regeneration. Lenzing is promoting the fibre for use in clothing worn next to the skin and in home furnishings like bed sheets [119].

Vitamin E is also known as "tocopherol" and belongs to the group of fat-soluble vitamins. It is available in nature from various fruit and vegetable oils. The oil has good antioxidant and moisture-binding properties. Human skin generates free radicals during exposure to the Sun and UV light, which may damage the skin. Tocopherol antioxidants act as radical scavengers, hence their use for skin ailments and for protecting skin cells from oxidative stress [120, 121]. The antioxidant properties of silk sericin have also been evaluated; radical scavenging activity was enhanced by 56 % in a treated sample compared with a control sample [46]. Natural products—such as sericin, chitosan and aloe vera—can be effectively used in the development of sustainable traditional as well as fashionable textiles, while imparting added value to such products.

9 Summary

Fashion can be encapsulated as the prevailing styles manifested by human behaviour and the latest creations by the designers of textile and clothing, footwear, body piercing, decor, etc. Fashion can trace its history to the Middle East (i.e., Persia, Turkey, India and China); it then gradually spread into Europe. Initially, fashionable products were only affordable by the royals or the rich; however, as civilization progressed, at least since the end of the 18th century, slowly such products became affordable to the middle classes and finally to the people of the world. With the evolution of society in the last few centuries, people have become more and more concerned about their lifestyle, fashion, health, hygiene, food and beverage, comfort, luxury, leisure and wellbeing. Until the 1950s, natural fibres—such as cotton, flax, ramie, hemp, wool, silk, pashmina and paul—were utilized in the production of fashionable textiles. Later on, many synthetic fibres—such as polyester, acrylic, spandex, viscose rayon, cellulose acetate and nylon—began to penetrate the fashion market due to advances in polymer science, material science and textile science, bringing substantial changes in fibre and fabric design and development, engineering new polymers for fibre formation, understanding and controlling fibre morphology, fibre characteristics and performance, their durability and interaction with the environment in greater depth. Fashionable textiles based on wholly natural or synthetic fibres as well as blends of both have been developed to impart the requisite fashion attributes during spinning, weaving, knitting, non-woven production and high-end chemical finishing. As some traditional textile-processing chemicals and auxiliaries have significant adverse effects on the environment, much research has taken place into sustainable fibre formation/application, dyeing and chemical finishing in attempts to mitigate issues such as global warming, environmental pollution and climate change. Since natural fibres are sustainable, they are gaining in importance for making green, ethical and sustainable eco-fashion owing to such properties as bio-degradability, renewablity, carbon neutrality, high-moisture regain, soft feel, adequate to fair strength and look good quality after chemical treatment. Naturally coloured cotton, organic cotton, organic wool, wild silk, flax and hemp are the important fibres and raw material for the sustainable fashion industry. Efforts have also been made in recent times for sustainable dyeing and value-added finishing of textiles using various plant/herbal extracts, bio-materials, bio-polymers and bio-molecules—such as enzymes, natural dyes, bio-mordants, aromatic and medicinal plants, chitosan, aloe vera, neem, lignin, silk sericin, grape and mulberry fruit extract, citrus oil, lemon oil and tulasi extract.

References

1. http://www.naturalfibres2009.org/en/iynf/sustainable.html, Dated 12-11-2015
2. Samanta KK, Basak S, Chattopadhyay SK (2015) Book chapter on sustainable UV protective apparel textile. In: Muthu SS (ed) Hand Book of sustainable apparel production. CRC Press Taylor and Francis, pp 113–137

3. Basak S, Samanta KK, Chattopadhyay S K, Saxena S, Parmar MS (2015) Spinach leaf (Spinacia Oleracea): a bio-source for making self-extinguishable cellulosic textile. Indian J Fibre Text Res (in press)
4. Basak S, Samanta KK, Saxena S, Chattopadhyay SK, Narkar R, Mahangade R (2015) Flame retardant cellulosic textile using bannana pseudostem sap. Int J Cloth Sci Technol 27(2):247–261
5. http://www.fashionmegreen.com/what-are-sustainable-fibres, Dated 12-11-2015
6. Wannajun S, Srihanam P (2012) Development of thai textile products from bamboo fibre fabrics dyed with natural indigo. Asian J Text 2(3):44–50
7. https://en.wikipedia.org/wiki/Sustainable_fashion, Dated on 12-11-2015
8. http://www.triplepundit.com/special/sustainable-fashion-2014/rise-sustainable-fibers-fashion-industry/, Dated on 12-11-2015
9. http://www.naturalfibres2009.org/en/fibres/hemp.html, Dated 12-11-2015
10. Ammayappan L, Nayak LK, Ray DP, Das S, Roy AK (2013) Functional finishing of jute textiles—an overview in india. J Nat Fibers 10:390–413
11. Ahmed Z, Akhter F, Hussain MA, Haque MS, Sayeed MMA, Quashem MA (2002) Research on jute and allied fibre plants. Pak J Biol Sci 5(7):812–818
12. Roy AN, Basu G (2010) Development of newer products with spun wrapped Jute yarns. Indian J Nat Prod Res 1(1):11–16
13. Sengupta S, Debnath S (2012) Studies on jute based ternary blended yarns. Indian J Fibre Text Res 37:217–223
14. Sharma BK (2008) A book on "ramie: the steel wire fibre. DB Publication, Guwahati, India, pp 1–143
15. Data Book on fibres allied to jute (2012) Director-National Institute of Research on Jute and Allied Fibre Technology, pp 1–70
16. Basak S, Samanta KK, Chattopadhyay SK, Narkar R (2015) Self-extinguishable lingo-cellulosic fabric using Banana Pseudostem Sap. Curr Sci 108(3):372–383
17. Roy DP, Bhaduri SK, Nayak LK, Ammayappan L, Manna K, Das K (2012) Utilization and value addition of banana fibre-A review. Agri Rev 33(1):46–53
18. Basu G, Roy AN (2008) Blending of jute with different natural fibres. J Nat Fibers 4(4):13–29
19. Majeed K, Jawaid M, Hassan A, Bakar AA, Khalil HPSA, Salema AA, Inuwa I (2013) Potential materials for food packaging from nanoclay/natural fibres filled hybrid composites. Mater Des 46:391–410
20. Roy AN, Basu G, Pan NC (2015) Processing of banana fibre in jute spinning system and product development. Indian J Nat Fibres 1(2):185–193
21. https://en.wikipedia.org/wiki/Bast_fibre, Dated 12-11-2015
22. Amao IO, Adebisi-Adelani O, Olajide-Taiwo FB, Adeoye IB, Bamimore KK, Olabode I (2011) Economic analysis of pineapple marketing in Edo and Delta states in Nigeria. Libyan Agric Res Center J Int 2:205–208
23. Sengupta S, Debnath S (2010) A new approach for jute industry to produce fancy blended yarn for upholstery. J Sci Ind Res 69:961–965
24. Roy AN, Basu G (2010) Improvement of a traditional knowledge by development of jacquard shedding based handloom for weaving ornamental jute fabric. Indian J Tradit Knowl 9(3):585–590
25. Roy AN, Basu G, Bhattacharya GK (2009) An approach to engineer jute yarn for improvement of its property performance. J Inst Engineers (India) Text Engg Div 89:3–9
26. Mazunder MC, Sen SK, Dasgupta PC (1975) Indian Text J 85(8):135
27. Doraiswami I, Chellamani P (1993) Jute/cotton blends. Asian Text J 1(8):53–56
28. Azad MAK, Sayeed MMA, Kabir SMG, Khan AH, Rahman SMB (2007) Studies on the physical properties of jute-cotton blended curtain and 100 % cotton curtain. J Appl Sci 7 (12):1643–1646
29. Roy AN, Basu G, Majumder A (2000) A study on warp-spun jute yarn with cellulosic yarn as wrapping element. Indian J Fibre Text Res 25:92–96

30. Chollakup R, Tantatherdtam R, Ujjin S, Sriroth K (2010) Pineapple leaf fiber reinforced thermoplastic composites: Effects of fiber length and fiber content on their characteristics. J Appl Polym Sci 119:1952–1960

31. Sinha MK (1974) Rope making with banana-plant fibre. J Text Inst 65:612–615

32. Sinha MK (1974) The use of banana-plant fibre as a substitute for jute. J Text Inst 65:27–33

33. Aditya RN, Ganguli AK, Som NC (1981) Development of jute based yarns for carpets. Indian Text J 91(4):77–84

34. Ganguli AK, Aditya RN, Som NC (1980) Development of products from blends of jute and natural and synthetic fibres on jute processing system—I. Man-made Text India 23(7):317–331

35. Ganguli AK, Aditya RN, Som NC (1980) Development of products from blends of jute and natural and synthetic fibres on jute processing system—II. Man-made Text India 23(8):410–417

36. Ganguly PK, Sengupta S, Samajpati S (1999) Mechanical behaviour of jute and polypropylene blended needle-punched fabrics. Indian J Fibre Text Res 24:34–40

37. Bhardwaj S, Juneja S (2012) Performance of jute viscose/polyester and cotton blended: yarns for apparel use. Stud Home Com Sci 6(1):33–38

38. Ammayappan L (2013) Eco-friendly surface modification of wool fibre for its improved functionality: an overview. Asian J Text 3(1):15–28

39. Shakyawar DB, Raja ASM, Kumar A, Pareek PK (2015) Antimoth finishing treatment for woolens using tannin containing natural dyes. Indian J Fibre Text Res 40:200–202

40. Ammayappan L, Moses JJ (2011) Study on improvement in handle properties of wool/cotton union fabric by enzyme treatment and subsequent polysiloxane-based combination finishing. Asian J Text 1(1):1–13

41. Aramwit P, Siritientong T, Srichana T (2012) Potential applications of silk sericin, a natural protein from textile industry by-products. Waste Manage Res 30(3):217–224

42. Ki CS, Kim JW, Oh HJ, Lee KH, Park YH (2007) The effect of residual silk sericin on the structure and mechanical property of regenerated silk filament. Int J Biol Macromol 41:346–353

43. Poza P, Perez-Rigueiro J, Elices M, Lorca J (2002) Fractographic analysis of silkworm and spider silk. Eng Fract Mech 69:1035–1048

44. Teli MD, Rane VM (2011) Comparative study of the degumming of Mulberry, Muga, Tasar and Ericream silk. Fibres Text East Eur 19:1014

45. Mondal M, Trivedy K, Kumar SN (2007) The silk proteins, sericin and fibroin in silkworm, Bombyx mori Linn.-a review. Caspian J Environ Sci 5:63–76

46. Gulrajani ML (2008) Bio-and nanotechnology in the processing of silk. http://www. fibre2fashion.com/industry-article/pdffiles/16/1517.pdf. Published November 19, 2008, Downloaded 11-03-2015

47. Gupta D, Chaudhary H, Gupta C (2014) Sericin-based polyester textile for medical applications. J Text Inst 105(5):1–11

48. Ammayappan L, Shakyawar DB, Krofa D, Pareek PK, Basu G (2011) Value addition of pashmina products: present status and future perspectives—a review. Agri Rev 32(2):91–101

49. Raja ASM, Shakyawar DB, Pareek PK, Temani P, Sofi AH (2013) A novel chemical finishing process for cashmere/pva-blended yarn-made cashmere fabric. J Nat Fibers 10:381–389

50. Shakyawar DP, Raja ASM, Kumar A, Pareek PK, Wani SA (2013) Pashmina fibre-Production, characteristic and utilization. Indian J Fibre Text Res 28:207–214

51. Ammayappan L (2014) Finishing of angora rabbit fibres. Am J Mater Eng Technol 2(2):20–25

52. Chattopadhyay SK, Bhaskar P, Ahmed M, Gupta NP, Plkharna AK (2005) Properties of indigenous angora rabbit hair and cotton blended yarns using short staple cotton spinning system. Indian J Fibre Text Res 30(5):215–217

53. Dirgar E, Oral O (2014) Yarn and fabric production from angora rabbit fiber and its end-uses. Am J Mater Eng Technol 2(2):26–28

54. Danzan B, Tsedev K, Luvsandorj N (2014) The shedding and growth dynamics of yak down wool and links to habitat ecological condition. Asian J Agric Rural Dev 4(2):156–161
55. http://webcache.googleusercontent.com/search?q=cache: http://180.211.172.109/ ifost2014Pro/pdf/S6-P281.pdf&gws_rd=cr&ei=cjBEVvv-OcG90gTf35z4CQ, Dated on 12-11-2015
56. https://en.wikipedia.org/wiki/Yak#Yak_fiber, Dated 12-11-2015
57. www.rangelands.org/internationalaffairs/2012_Symposia/pdf/Mongolian%20Yak% 20Industry.pdf, Dated on 12-11-2015
58. Liu J, Hu Y, Yu W (2009) The retraction investigation of yak hair fiber during roller stretching. J Fiber Bioeng Inform 1(4). doi:10.3993/jfbi03200905]
59. https://en.wikipedia.org/wiki/Yak_fiber, Dated on 12-11-2015
60. Raja ASM, Shakyawar DB, Pareek PK, Wani SA (2011) Production and performance of pure cashmere shawl fabric using machine spun yarn by nylon dissolution process. Indian J Small Ruminant 17:203–206
61. Chattopadhyay SK, Chattopadhyay AK, Ahmed M, Gupta NP, Pokharna AK (2001) Utilisation of Angora rabbit hair in blended with cotton for value-added fabrics. Asian Text J 10(3):86–91
62. Shah SA, Paralkar N, Chattopadhyay SK, Ahmed M, Gupta NP (2004) Some Aspects of processing wool/cotton & Angora rabbit hair/cotton blended fabrics. Man-made Text India 47(5):160–162
63. Samanta KK, Jassal M, Agrawal AK (2010) Antistatic effect of atmospheric pressure glow discharge cold plasma treatment on textile substrates. Fibre Polym 11(3):431–437
64. Samanta KK, Jassal M, Agrawal AK (2006) Atmospheric pressure glow discharge plasma and its applications in textile. Indian J Fib Tex Res 31(1):83–98
65. Samanta KK, Gayatri TN, Shaikh AH, Saxena S, Arputharaj A, Basak S, Chattopadhyay SK (2013) Effect of helium-oxygen plasma treatment on physical and chemical properties of cotton textile. Int J Biores Sci 1(1):57–63
66. http://www.newsweek.com/fabrics-even-finer-cashmere-69831, Dated on 12-11-2015
67. http://www.globalnaturalfibres.org/yak_wool, Dated 12-11-2015
68. http://www.imrsheep.com/yak.html, Dated 12-11-2015
69. Samanta AK, Aggarwal P (2009) Application of natural dyes on textiles. Indian J Fibre Text Res 34:384–399
70. Chattopadhyay SN, Pan NC (2013) Development of natural dyed jute fabric with improved colour yield and UV protection characteristics. J Text Inst 104(8):808–818
71. Teli MD, Adivarekar RV, Bhagat M, Manjrekar SG (2002) Response of jute to the dyes of synthetic and natural origin. J Text Assoc 62:129–134
72. Pan NC, Chattopadhyay SN, Dey A (2003) Dyeing of jute with natural dyes. Indian J Fibre Text Res 28:339–342
73. Chattopadhyay SN, Pan NC, Roy AK, Khan A (2010) Finishing of jute fabric for value-added products. J of Nat Fibers 7:155–164
74. Annual Report of National Institute of Research on Jute and Allied Fibre Technology, pp 1–76 (2015) www.nirjaft.res.in, Dated 12-11-2015
75. Kumar A, Pareek PK, Raja ASM, Shakyawar DB (2015) Extraction from babul (*Acacia Nilotica*) bark and efficacy of natural colour on woollen yarn. Indian J Small Ruminants 21 (1):92–95
76. Samanta AK, Konar A, Chakrabarti S (2011) Dyeing of jute fabric with tissue extract: Part 1-effect of different mordants and dyeing process variables. Int J Fibre Text Res 36:63–73
77. Samanta AK, Konar A (2011 Nov) Dyeing of textiles with natural dyes. Intech publication, China
78. Rajendran R, Radhai R, Balakumar C (2012) Synthesis and characterization of neem chitosan nanocomposite for development of antimicrobial cotton textile. J Eng Fibre Fabr 7:46–49
79. Singh SV, Purohit MC (2012) Applications of eco-friendly natural dye on wool fibres using combination of natural and chemical mordants. Univ J Environ Res Technol 2:48–55

80. Mongkholrattanasit R, Krystafek J, Wiener J, Studnickova J (2011) Properties of wool and cotton fabrics dyed with Eucalyptus, Tannin and Flavonoids. Fibres Text Eastern Europe 19:90–95
81. www.csiro.au/Organisation-Structure/Flagships/.../Camouflage.aspx, Dated 12-11-2015
82. Zhang H, Zhang JC (2008) Near infrared green camouflage of cotton fabric using vat dyes. J Text I 99:83–88
83. Gondrazi U, Mokhtari J, Nouri M (2014) Camouflage of cotton fabrics in visible and NIR region using three selected vat dyes. Indus Chem 39:200–209
84. www.ilovetocreate.com/ProjectDetails.aspx?name=Cool...Camo...Dye..., Dated 12-11-2015
85. www.science.howstuffworks.com/invisibility-cloak4.htm, Dated 12-11-2015
86. Apodaca JK (1990) Naturally coloured cotton: a new niche in the Texas natural fibres market. Working Paper series, Bureau of Business Research Paper number 1990-2
87. Chattopadhyay SK, Shanmugam N, Upadhye DL, Chaphekar AK, Krishna Iyer KR (2001) Spinning and fabric forming trials on naturally coloured cottons—some observations. Asian Text J 10(1):50–56
88. James M, Vreeland Jr (1999 Apr) The Revival of colored cotton. Scientific American 280 (4):112
89. Special Issue-National Seminar on Eco-friendly Cotton (1996 Sep) J Indian Soc Cotton Improve 21(2)
90. Parmar MS, Sharma RP (2002) Development of various colours and shades in naturally coloured cotton fabrics. Indian J Fibre Text Res 27:397–407
91. Werber FX (1994) Agriculture research service, USDA. Personal Communication: 1-31-94
92. Williams B (1994) Foxfibre naturally coloured cotton, green and brown (Coyote): Resistance to change in colour when exposed to selected stains and fabric care chemicals. University, Texas Tch
93. http://www.researchgate.net/...microencapsulation.../543fca210cf2be1758cfd4, Dated 12-11-2015
94. www.cancer.gov/cancertopics/pdq/cam/aromatherapy/.../page5, Dated 12-11-2015
95. Specos MMM, Escobar G, Marino P, Puggia C, Victoria M, Tesoriero D, Hermida L (2010) Aroma finishing of cotton fabrics by means of microencapsulation techniques. J Ind Text 40 (1):13–32
96. Samanta KK, Basak S, Chattopadhyay SK (2014) Eco-friendly coloration and functionalization of textile using plant extracts. In: Muthu SS (ed) Roadmap to sustainable textiles and clothing: environmental and social aspects of textiles and clothing supply chain. Springer, pp 263–287
97. Samanta KK, Basak S, Chattopadhyay SK (2015) Speciality chemical finishes for sustainable luxurious textiles. In: Gardetti MA, Muthu SS (eds) Handbook of sustainable luxury textiles and fashion. Springer 145–184
98. www.clevertex.cz/en/seaweed-in-textiles-wellness, Dated 12-11-2015
99. www.discoverspas.com/news/newsproducts145.shtml, Dated 12-11-2015
100. www.indiamart.com/princekatariatextiles/, Dated 12-11-2015
101. Elshafei A, El-Zanfaly HT (2011) Application of antimicrobials in the development of textiles. Asian J Appl Sci 4(6):585–595
102. Thilagavati G, Rajendrakumar K, Rajendran R (2005) Development of eco-friendly antimicrobial textile finishes using herbs. Indian J Fibre 30:431–436
103. Rajendran R, Radhai R, Balakumar C, Hasabo A, Mohammad A, Vigneshwaran C, Vaideki K (2012) Synthesis and characterization of neem chitosan nanocomposites for development of antimicrobial cotton textile. J Eng Fiber Fabr 7:136–141
104. Joshi M, Ali SW, Rajendran S (2007) Antibacterial finishing of polyester cotton blend fabric using neem: a natural bioactive agent. J Appl Polym Sci 106:785–793
105. Ahmed HA, Rajendran R, Balakumar C (2012) Nanoherbal coating of cotton fabric to enhance antimicrobial durability. Elixir Appl Chem 45:7840–7843

106. Sathianarayanan MP, Bhatt NV, Kokate SS, Walung VL (2010) Antibacterial finishing of cotton fabrics from herbal products. Int J Fibre Text Res 35:50–58
107. Varghese J, Tumkur VK, Ballal V (2013) Antimicrobial effect of Anecordiumoccidentale leaf extract against pathogens causing periodical disease. Adv Biosci Biotechnol 4:15–18
108. Joshi M, Ali SW, Purwar R (2009) Ecofriendly antimicrobial finishing of textiles using bioactive agents based on natural products. Int J Fibre Text Res 34:295–304
109. Gupta D, Laha A (2007) Antimicrobial activity of the cotton fabric treated with Quercusinfectoria extract. Indian J Fibre Text Res 32:88–92
110. Ammayappan L, Moses JJ (2009) Study of antimicrobial activity of aloevera, chitosan, and curcumin on cotton, wool, and rabbit hair. Fibers Polym 10(2):161–166
111. Raja ASM, Thilagavathi G (2011) Influence of enzyme and mordant treatment on the antimicrobial efficacy on natural dyes on wool material. Asian J Text 1(3):138–144
112. Basu SN, Ghose R (1962) A microscopical study on the degradation of jute fiber by micro-organisms. Text Res J 32(8):677–694
113. www.discovery.org.in/PDF_Files/de_031002.pdf, Dated 12-11-2015
114. www.ril.com/downloads/pdf/tech_publications.pdf, Dated 12-11-2015
115. http://assets.babycenter.com/ims/advertorials/IN/JB_wipes/home, Dated 12-11-2015
116. www.karawipes.com/skincarewipes_deepporecleansing, Dated 12-11-2015
117. www.iherb.com/United-Exchange...Jojoba-Oil-Aloe-Vera...1.../54680, Dated 12-11-2015
118. Samanta KK, Basak S, Chattopadhyay SK (2015) Recycled fibrous and non-fibrous biomass for value added textile and non-textile applications. In: Muthu SS (ed) Environmental implication of recycling and recycled products. Springer, pp 167–212
119. www.textileworld.com/Issues/2011/MarchApril/Quality_Fabric_Of_The_Month/Textile_ Cosmetics, Dated 12-11-2015
120. www.ncbi.nlm.nih.gov › NCBI › Literature › PubMed Central (PMC), Dated 12-11-2015
121. Cohen TD, Koren R, Liberman A, Ravid A (2006) Vitamin D protects keratinocytes from apoptosis induced by osmotic shock, oxidative stress, and tumor necrosis factor. Ann NY Acad Sci 43:350–353

Milkweed—A Potential Sustainable Natural Fibre Crop

T. Karthik and R. Murugan

Abstract The demand for renewable raw materials is steadily rising as the drive for a green economy and a sustainable future accelerates. Mounting environmental issues and changing attitudes of consumers have made petroleum-based manufactured products more expensive and less desirable in the present world (as of 2016). Utilization of less valuable lignocellulosic fibres that are abundant and do not require irrigation for their cultivation makes them inexpensive and bio-degradable alternatives to petroleum-based synthetic fibres. These fibres are found in dry habitats in the USA, Asia and South Africa and contain over 2000 species worldwide, the properties of which slightly differ as a result of different soil conditions. In recent years, much research has been carried out into potential applications of milkweed fibre due to its incredible characteristics—such as hollowness, low density and hydrophobicity—particularly in technical textiles. The milkweed plant can adapt to almost any soil condition from swampy and moist to sandy and arid. It is a perennial plant and hence, once planted, does not require replanting each season and does not require any fertilizers, making the plant sustainable. Successful commercialization of milkweed as a crop is dependent upon mechanized harvesting, handling, drying and floss-processing systems. This chapter gives comprehensive information about the history of the milkweed plant and fibres, fibre morphology and characteristics, spinnability of milkweed fibres, fabric properties and potential application of milkweed fibres in such areas as clothing, lightweight composites, oil sorption as well as thermal and acoustic insulation.

1 Introduction

According to the UN World Commission on Environment and Development (WCED)—also known as the Brundtland Report—sustainability is defined as "meeting the needs of the present without compromising the ability of future

T. Karthik (✉) · R. Murugan
Department of Textile Technology, PSG College of Technology, Coimbatore, India
e-mail: peacesoull23@gmail.com

© Springer Science+Business Media Singapore 2016   111
S.S. Muthu and M.A. Gardetti (eds.), *Sustainable Fibres for Fashion Industry*,
Environmental Footprints and Eco-design of Products and Processes,
DOI 10.1007/978-981-10-0566-4_6

generations to meet their own needs" [1]. The UK's Forum for the Future defined it in 2006 as "A dynamic process which enables all people to realize their potential and improve their quality of life in ways which simultaneously protect and enhance the Earth's life support system". In simple words, sustainability is development for environmental, economic and social wellbeing for today and tomorrow. It is based on two concepts: the need for and limitations to the environment's ability to meet present and future needs.

Renewable resources, which were important for the wealth of people before the Industrial Revolution, are today gaining more interest as a result of their positive effects on agriculture, environment and economy. Renewable resources have significant advantages over synthetic resources; they contribute to the conservation of finite fossil resources and add nothing to the greenhouse effect. Natural fibres are valuable in that they compete and co-exist with synthetic fibres in terms of quality, sustainability and economy of production. They are generally characterized by good air permeability, hygroscopicity, moisture absorption and bio-degradability [2]. Some natural fibres also possess unique properties such as heat conduction, improved dyeability, resistance to mildew/bacteria, flame retardancy and UV-blocking ability. After World War II the increase in synthetic fibre production greatly reduced the use of natural fibres.

The demand for sustainable and renewable raw materials is progressively growing as the drive for a green economy and a sustainable future accelerates. Increasing environmental concerns and shifting consumer attitudes towards sustainability have made petroleum-based manufactured products more expensive and less desirable in today's world [3–5].

Extensively used sustainable raw materials include plant fibres (cotton, linen, hemp), animal fibres (wool, silk) and synthetic fibers (polyester, aramid, acrylic, nylon, spandex, carbon). The production of both natural/synthetic fibre polymers has a significant impact on the environment [6]. Cotton—the most used natural fibre when grown by conventional means—requires enormous amounts of pesticides and water as well as large quantities of chemicals for processing and dyeing. Regenerated cellulosic fibre—such as rayon and viscose—are made of cellulose from trees but need chemicals so that it can be processed to become a useful polymer. Other synthetic fibres—such as polyester, nylon and acrylics—rely on non-renewable petroleum sources for production. Thus, such polymers impact the environment in a major way in terms of depletion of fossil fuels, increased number of landfills, dumping of waste in the ocean, increased emission of CO_2, pollution caused by toxic emissions and recycling of plastic. All of this has a negative effect on ecosystems and contributes to increased global warming [7–9]. From the economic point of view, dwindling oil supplies are likely to boost oil prices while energy costs skyrocket (although oil prices are currently in freefall). Furthermore, processing of these fibres to achieve the end product involves environmental and health risks.

Efforts to find a fibre as a substitute for natural and man-made fibres are crucial to having an adequate supply of fibres at a reasonable cost in future. The scarcity of petroleum resources—the raw material for synthetic fibres—and diminution in the availability of water, land and other resources to cultivate natural fibres could lead to increase in the cost of fibres and, hence, garments in the near future [10]. Further, higher income for farmers from bio-fuel crops such as corn is leading to a decline in the production of natural fibres, especially cotton—hence the search for sustainable, abundantly available, and renewable lignocellulosic by-products [11, 12]. In this respect, several lignocellulosic fibres—such as jute, ramie (flowering plant in the nettle family), sisal, hemp and coir—have been explored as potential applications [13, 14]. There exist a number of vegetable fibres that have not been investigated to any great extent; they remain unutilized and are dumped as natural waste. This chapter provides comprehensive information about one such natural cellulosic fibre —milkweed. The cultivation process, properties and structure of fibres, spinnability and their application in various fields are discussed in detail.

2 Weed Versus Cotton

A weed can be defined as [15]:

- A plant out of place and are not intentionally sown
- A plant grown at unwanted place
- A plant whose qualities not yet fully exposed
- Plants which are competitive, persistent, pernicious, and interfere negatively with human activity and many others.

Weeds are quite simply plants that have evolved naturally without any input from humans. Humans and nature are involved in the breeding of wild plants but the key difference between the two is that humans breed plants for yield, while nature breeds plants for existence. Generally, weeds have two important advantages over cultivated crops like cotton [16]: they can grow in poor soil without the need for irrigation and fertilizer and they are more resistant to pests (hence pesticides are not required). The general benefits of weed plants are:

- Soil stabilization
- Habitat and food for wildlife
- Nectar for bees
- Aesthetic qualities
- Addition of organic matter
- Provision of a genetic reservoir
- Human consumption
- Provision of employment opportunities.

3 History of Milkweed

Asclepias, commonly denoted as milkweed, is a genus of herbaceous perennial, dicotyledonous plants comprising more than 140 known species worldwide, 108 of which are native to North America. The Latin name comes from Asclepius, the Greek god of medicine and healing (Fig. 1), because of its several applications in phytotherapy, an alternative medicine based on natural extracts and health-promoting agents.

The cultivation of milkweed plants (mainly to attract butterflies) in gardens could degrade the environment and agricultural land as well as threaten biodiversity should they escape back into the wild, and their removal can cause expense for farmers and land owners [17]. One such species is common milkweed (*Asclepias syriaca* L.). *Asclepias* spp are generally considered as being persistent, hardy weeds consisting of cardiac glycosides that are toxic when consumed by livestock [18]. Milkweed fibre was used in ancient times in North America and Africa, two areas where many species of *Asclepias* are indigenous. Milkweed have been of interest to agriculturalists for several years owing to its potential economic value as a new crop [19–21].

The fragrant flowers, latex and hard roots of milkweed plants attracted a great deal attention not only among researchers but also agriculturists. A Frenchman, Louis Hebert, was the first to send milkweed seeds from New France (Canada) to Paris (France); plants were later studied by botanist-cum-medical doctor Philip Cornut [22–24]. Gleditsch from Germany was the first to utilize milkweed fibre in 1746 for padding and later for cloth and lustrous velvets. The limitations of milkweed—such as poor germination of seed in cultivation, resistance to dyeing and weakness and brittleness of fibres for spinning—were summarized by Lichtenstein [25].

Fig. 1 Stone carving of Asclepius, Greek god of medicine (*Source* www.scoliosisjournal.com)

In Germany and America a great deal of research was carried out into the application, cultivation and harvesting of milkweed fibre. During World War I, Schuroff from Germany concluded that common milkweed (*Asclepias syriaca*) could not be adapted as a domestic source of fibre. However, [26] revealed that it could be a potential substitute for kapok in upholstery.

During World War II, milkweed floss was considered tactically and economically significant, with more than a million pounds (453,593 kg) of milkweed seed fibre being used to fill life preservers and other flotation equipment [19, 22]. It was mainly considered for stuffing insulating material. The scarcity of kapok fibre during the war was a major reason for the demand for milkweed fibre during this period. Milkweed floss is made of fibres with a large lumen and very thin walls that grant kapok fibre its elasticity. The fibres also have a waxy coating that makes them hydrophobic. Owing to its lightweight, hydrophobic and insulating properties, milkweed floss was identified as an ideal kapok substitute [17, 27, 28].

Two types of fibre can be obtained from the milkweed plant: the long, strong but brittle bast fibre and the seed hair fibre known as floss. Even though bast fibres are similar to flax in many aspects, fabrics from bast fibres are very brittle with little or no draping qualities. Though pulp produced from the fibre yielded good-quality paper, it proved too costly for economic utilization [29, 30].

Milkweed was investigated as a potential biofuel during the 1970s and 1980s in the USA, but unfortunately was found to be uneconomical [31–33]. Further, milkweed has been investigated as a substitute for crude petroleum [34] and milkweed latex has been investigated for the production of natural rubber [22, 35]. Today, milkweed floss is used commercially as a hypoallergenic filling for pillows and continental quilts and milkweed seed oil is utilized in novel bodycare products [36] (Ogallala Comfort Co. 2014).

Since milkweed plants are widely grown, they have been given a number of names—such as swallow wort, dead sea apple, desert wick, mudar (Indian name), estabragh (Persian name) and rubber bush. Milkweed plants growing naturally in the wild state are normally found in continental regions under sandy and arid (hot and dry) to moist and swampy (temperate and humid) conditions, since they easily adapt to any soil conditions [37]. Hence, the easy soil adaptability and weed-like nature of milkweeds make them cost effective.

4   Milkweed Plant Morphology

Milkweed, a perennial plant that can adapt to adverse soil conditions, has been considered as an alternative source of fibre in recent years. It formerly belonged to the family Asclepiadaceae, but is now categorized into the subfamily Asclepiadoideae of the dogbane family Apocynaceae. Farmers and scientists joined hands in the late 1980s to develop milkweed as an alternative fibre source [17, 27, 31, 36, 38, 39].

Milkweed got its name from the milky liquid it produces, which comprises latex and a few complex chemicals that make the plants indigestible to most animals. When wounded, the stems, leaves, and pods of most milkweed plants ooze this liquid—an exception is butterfly milkweed (*Asclepias tuberosa*) [40].

Milkweed flowers are normally organized in rounded flower clusters and the stalks supporting these flowers are either erect or loose based on the species or variety. The upper part of each flower, the corona, consists of five hoods. The corolla, comprising five petals fused together, forms the lower part of the flower [32, 41, 42]. The shape of the hood varies significantly with the species and some have horn-like attachments (Fig. 2). Flower colour varies within the genus and includes various shades of white, yellow, green, purple, pink, orange and red.

Milkweeds produce their seeds in follicles. The seeds, which are arranged in overlapping rows, have white silky filament-like hairs known as "pappus", "comas" or "floss". The follicles ripen, split open and the seeds, each carried by dried floss, are blown by the wind [43]. Aquatic milkweed (*Asclepias perennis*), which is native to the USA, is the only species without floss (Fig. 3); it is adapted for water dispersal [44]. The different stages of growth of milkweed fibres are shown in Fig. 4. Flowering may occur each year between August and January in India, with fruits maturing from October to February [24].

Milkweed plants can be stout, with erect stems, and grow up to 6 feet tall (1.82 m) such as common milkweed and showy milkweed (*Asclepias speciosa*), while a few species are low growing and sprawling such as pinewood milkweed (*Asclepias humistrata*) and pallid milkweed (Asclepias cryptoceras). Plant growth is variable depending on the species. Even within a species, plant height can vary greatly depending on local genetics and climate. The root morphology of milkweed plants ranges from fleshy to woody and many species are very deep rooted [45].

In addition to reproducing by seed, some milkweeds reproduce vegetatively by producing new shoots from adventitious buds on their roots [46]. Shoots can even emerge from lateral roots located several feet away from the aboveground growth of the parent plant. The list of possible by-products is long. Every part of the milkweed plant is capable of yielding a variety of useful materials (as shown in Fig. 5) [45]. The composition of milkweed pods is shown in Fig. 6.

Fig. 2 Different components of the milkweed flower (Reproduced with permission from [32])

 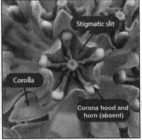

Fig. 3 Aquatic milkweed seeds without floss (Reproduced with permission from [32])

Fig. 4 Various stages of milkweed fibre growth: **a** plant with flowers; **b** flowering stage; **c** development of pods; **d** mature pods with seeds; **e** open and split mature pods; **f** fibres with seed (Reproduced with permission from [2])

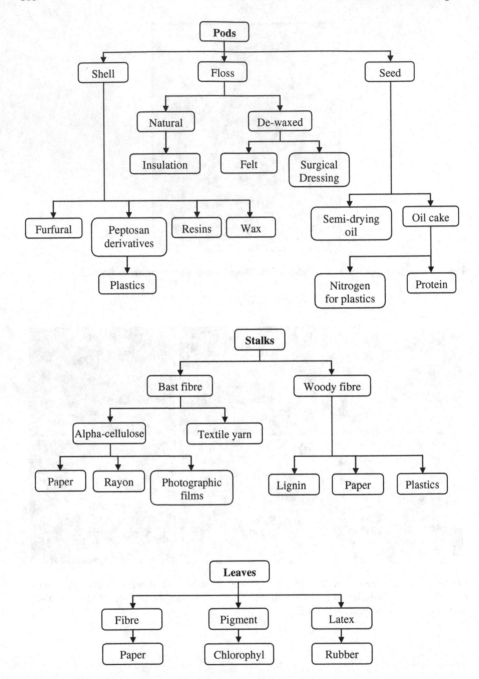

Fig. 5 Materials produced by the milkweed plant (*Source* [46])

Fig. 6 Composition of a
mature milkweed pod
(*Source* [2])

5 Milkweed Fibre Structure and Morphology

5.1 Fibre Surface Morphology

Like cotton, milkweed fibres are cellulosic growing on single cells in large seeds of
the plant. But, it does not have a convoluted structure like cotton. The fibres are
hollow. Although milkweed fibre does not collapse upon drying despite having a
low cell wall thickness, it does collapse during chemical treatments [30, 47–49].
Figure 7 shows longitudinal and cross-sectional scanning electron microscope
(SEM) images of milkweed fibres. The figure shows that these fibres have a
smooth, uniform and lustrous surface and that they are hollow.

The low-density and lightweight features of milkweed fibres are down to its
hollow structure and its very thin wall structure compared with its diameter. Owing
to its hollow structure, they possess good insulation or buoyancy properties which
make them suitable for filler fibres in comforters, life vests and winter jackets
[50–53]. The wall structure of fibres comprises three distinct regions—the inner

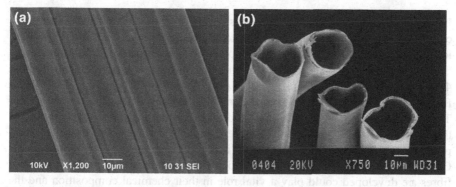

Fig. 7 Longitudinal and cross-sectional views of milkweed fibre (Reproduced with permission
from [2])

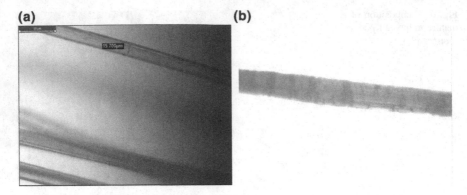

Fig. 8 Optical microscope photograph of milkweed fibre

wall, the outer wall and the microfibrils which are arranged in between [54]. Drean et al. [47] analyzed milkweed fibre wall thickness using an image analysis technique and observed the inner and outer diameters of the fibres using SEM. They found a wall thickness of around 1.27 μm but there was huge variation—a coefficient of variation (CV) of 30 %. Milkweed fibre wall thickness depends on whether the fibres are extracted from the stem or from the seed pod. The wall thickness varies from 6 to 7 μm and the aspect ratio (ratio between length and diameter, L/D) of milkweed stem fibres is approximately 67 % higher than milkweed seed fibres [29].

Milkweed fibre has also been analyzed under an optical microscope to understand its wall thickness and hollow structure. Optical microscope photographs of milkweed fibres are shown in Fig. 8. The longitudinal grooves could mean the fibres have excellent capillary effects, hygroscopicity and air permeability.

5.2 Chemical Composition of Milkweed Fibre

In general, lignocellulosic fibres consist of several components other than cellulose—such as hemicelluloses, holocellulose, lignin, wax and ash. The chemical composition of milkweed fibre compared with cotton is given in Table 1.

Most lignocellulosic agricultural by-products have a cellulose content of about 40–45 % [13], but the cellulose content of milkweed fibre is higher. It is evident from Table 1 that the chemical composition of different species of milkweed varies. The lignin content in stem fibres and seed fibres is quite high; this is the reason for the increased stiffness and brittle nature of milkweed fibres. The chemical composition of milkweed seed floss and its agronomic aspects are analysed by Campbell [7]. He reported that the environment in which populations of milkweed fibres are developed could play a vital role in their chemical composition and the agronomic performance of their progeny. Knudsen [55] stated that milkweed seed

Table 1 Chemical composition of milkweed fibre

Serial no.	Chemical composition (%)	Milkweed stem fibre [29]	Milkweed floss			Cotton
			Karthik and Murugan [49]	Ashori and Bahreini [29]	Knudsen [55]	
1	Cellulose	57 ± 3.2	59 ± 2.0	49 ± 1.1	55 %	94.0
2	Hemicellulose	19 ± 1.9	23 ± 1.0	20 ± 2.3	24 %	2.5
3	Lignin	18 ± 0.9	13 ± 0.8	23 ± 0.6	18 %	0.0
4	Ash	2.5 ± 0.4	1.5 ± 0.4	3.8 ± 0.4	–	1.2
5	Extractives	12 ± 1.4	4 ± 0.5	10 ± 0.9	3 %	0.9

fibre has a potential application as a superabsorbent fibre, thermal insulation fibre, fluid carrier fibre, bulking fibre, bonding fibre and touch-sensitive fibre.

The composition, structure and properties of common milkweed stem fibres were investigated by Reddy and Yang [56]. They found that they comprise 75 % cellulose, which is higher than milkweed floss but lower than cotton and linen. The crystallinity of milkweed stem fibres is lower than that of cotton or linen. Fibre strength is similar to cotton and fibre elongation is greater than that of linen. Overall, milkweed stem fibers have the requisite properties for high-value textiles, composites and other industrial applications.

5.3 Molecular Properties of Milkweed Fibre

Fibre structure and fibre cross-section are the two main parameters influencing the physical properties of fibre. Fibre structure is dependent on the size, shape and position of the cellulose molecules, the masses of molecules either as crystallites or fibrils as well as the region of amorphous cellulose [57]. The molecular properties of common milkweed fibres regarding alpha-cellulose content were investigated by Timell and Snyder [58], who found that the alpha-cellulose fraction was wholly composed of anhydro-glucose units and the hemicellulose portion of anhydro-oxylose units. In addition to the original material, it also contained minor amounts of arabinose and uronic acid residues. They also found that milkweed fibre had a rather interesting composition, apparently comprising only two main constituents— an exceedingly high-molecular-weight cellulose part and a probably low-molecular-weight hemicellulose portion, the latter consisting chiefly of xylan. The lower DP limit was 2500 and the upper was 8000, with a maximum located at a DP of 4000. Despite the presence of high-DP cellulose the fibre has low strength due to the high content of xylan.

Barth and Timell [59] analysed the constitution of hemicellulose from common milkweed floss. The average degree of polymerization of methylated hemicelluloses was 97 and the corresponding value for native polymers was 172. On the basis of these results, it is suggested that hemicelluloses contain approximately 170 β-D-xylopyranose residues linked together by 1,4-glycosidic bonds, with on average one branching point present per molecule. Every 14th anhydro-oxylose unit carries a single terminal side chain of 4-O-methyl-D-glucuronic acid attached by an α-glycosidic bond to the 2-position of xylose residues.

5.4 Crystallinity of Milkweed Fibre

The crystalline percentage of milkweed stem fibres compared with that of cotton (Fig. 9) was investigated by Reddy and Yang [56], who found that the crystalline percentage of milkweed stem fibres (39 %) was lower than that of cotton (65–70 %). The X-ray diffraction (XRD) results in Fig. 9 show two main peaks, representing the planes (002) and (101) at 2θ around 22.46 and 16.3°, respectively, which are assigned to cellulose I. Peak intensity at 22.46° is said to represent the total intensity (crystalline + amorphous) of the material and the peak intensity at 16° corresponds to the amorphous material in the cellulose. The crystalline percentage of milkweed fibres along with that of rux (Calotropis gigantea) and cotton was analysed by Louis and Andrews (1987), who found that the degrees of crystallinity of de-waxed cotton, milkweed and rux fibres were 90.3, 73.6 and 76.7 %, respectively, and the degrees of crystallinity of native fibres (raw stock) were 89.3, 72.0 and 76.5 %, respectively. Karthik and Murugan [49] found that the crystalline percentage of Pergularia daemia (the trellis-vine, a milkweed species) was around 55 %. Milkweed fibres have a lower crystalline fraction due to the lower amount of cellulose than other lignocellulosic fibers: 75 % for sisal, 83 % for cotton, 63.5 % for jute, 72.4 % for ramie, 71 % for flax, and 63.5–82.2 % for hemp [13, 60].

Fig. 9 XRD pattern of cotton and milkweed stem fibres (Reproduced with permission from [56])

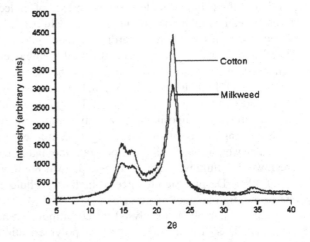

5.5 Chemical Resistance of Milkweed Fibre

The solubility of milkweed fibres when treated with various chemicals is given in Table 2. There is no damage when treated with volatile organic acids. There is a change in color when treated with strong alkalis, which fades even more if not controlled properly. The fibres were resistant to the action of acetic, formic, citric and dilute nitric acids as well as to sodium hydroxide and acetone (Table 2). On the other hand, fibres were readily soluble in hot and concentrated H_2SO_4, HCl and HNO_3 solutions.

5.6 Thermal Properties of Milkweed Fibre

Gu et al. [61] studied the two-stage thermal process (i.e., pyrolysis–combustion) to identify the major organic volatile products of pyrolysis and to correlate gas evolution with the decomposition of individual components of milkweed floss. During pyrolysis, acetic acid, formic acid and methanol are formed in addition to CO_2 and H_2O. The result showed that pyrolytic decomposition of the three chemical constituents of milkweed occur without any apparent synergistic interaction. The combustion of milkweed produced CO_2 and H_2O, but removal of the waxy coating from fibres resulted in increased susceptibility to combustion.

Further, they stated that the fibre undergoes three reactions subsequent to moisture loss: a small exotherm associated with major weight loss—58 % for milkweed and 62 % for extracted milkweed—between 100 and 300 °C; rapid combustion between 300 and 450 °C resulting in loss of most of the remaining sample mass (38 and 32 %, respectively); and a small weight loss recorded at temperatures between 450 and 600 °C (0.8 and 0.6 %). The thermal stability of milkweed floss is of the same order as its least stable component, hemicellulose. Cellulose has a higher degradation temperature and degradation onset temperature. Although lignin begins to degrade at a lower temperature than milkweed floss, it degrades more slowly as the temperature rises.

Karthik and Murugan [49] also analysed the thermal behaviour of milkweed fibre; the resultant differential scanning calorimetry thermogravimetric analysis (DSC-TGA) of the fibre is shown in Fig. 10. From the curves of raw milkweed fibres, three different regions were observed during thermal degradation of material. The first phase of weight loss started in the temperature range 30–110 °C, which was related to evaporation of water. The second major degradation occurred in the temperature range 180–420 °C, which could be related to the degradation of lignin and hemicellulose in the fibre. The last phase of weight loss occurred in the range 360–580 °C which indicated the degradation of alpha-cellulose and other non-cellulosic components of the fibre. The findings are in agreement with Dollimore and Holt (1973). The thermal degradation of cotton fibers generally occurs in three phases at temperature ranges of 37–150 °C, 225–425 °C and 425–600 °C, respectively [62].

Table 2 Solubility and reaction of milkweed fibre to various chemicals

Serial no.	Reagents	Milkweed fibre				
		Room temperature (22–25 °C)		Temperature of 40 °C		Boiling temperature
		Colour	Solubility	Colour	Solubility	Solubility
1	Dilute H$_2$SO$_4$ (10 %)	Dark cream	Insoluble	Dark cream	Insoluble	Insoluble
	Concentrated H$_2$SO$_4$ (98 %)	Greyish yellow	Soluble	Greyish yellow	Soluble	Dissolves completely
2	Dilute HCL (20 %)	Stained	Insoluble	Stained	Insoluble	Dissolves completely
	Concentrated HCL (38 %)	Pale pink	Partially soluble	Pale pink	Partially soluble	Insoluble
3	Dilute HNO$_3$ (10 %)	Dark green	No effect	Dark green	No effect	Insoluble
	Concentrated HNO$_3$ (90 %)	White to brown	Disintegrates	White to brown	Disintegrates	Color change to yellow, disintegrates
4	Dilute formic acid (10 %)	Cream	Insoluble	Cream	Insoluble	Insoluble
	Concentrated formic acid (85 %)	Cream	Insoluble	Cream	Insoluble	Insoluble
5	Dilute citric acid (10 %)	Cream	Insoluble	Cream	Insoluble	Insoluble
	Concentrated citric acid (80 %)	Cream	Insoluble	Cream	Insoluble	Insoluble
6	Dilute acetic acid (10 %)	White	Insoluble	White	Insoluble	Insoluble
	Concentrated acetic acid (99 %)	White	Insoluble	White	Insoluble	Insoluble
7	Dilute oxalic acid (10 %)	Cream	Insoluble	Cream	Insoluble	Insoluble
	Concentrated oxalic acid (80 %)	Cream	Insoluble	Cream	Insoluble	Insoluble
8	NaOH (18 %)	No color change	Insoluble	Colour change to yellow	Insoluble	White to greenish yellow, insoluble
9	Sodium carbonate (18 %)	No colour change	Insoluble	No colour change	Insoluble	White to yellowish brown, insoluble

Fig. 10 DSC-TGA curves of milkweed fibre

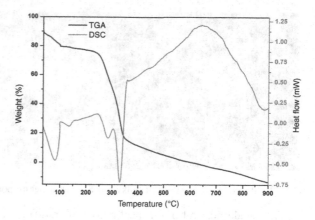

6 Physical Properties of Milkweed Fibre

The general physical properties of milkweed fibre analysed by various researchers are given in Table 3.

Table 3 Physical properties of milkweed fibre compared with those of cotton

Serial no.	Sample	Cotton	Milkweed species		
			Common milkweed [30]	*Calotropis procera* [64]	*Pergularia daemia* [2]
1	2.5 % span length (mm)	28.3	29.7	27.9	29.09
2	Uniformity ratio (%)	51.2	35.7	48.8	41.6
3	Strength (g/tex)	21.3	16.0	24.3	20.5
4	Elongation (%)	6.3	1.5	2.0 %	3.9
5	Micronaire (µg/in.)	3.5	<2.4	<2.4	<2.4
6	SFC (*n*)	25.2	–	–	33.3
7	Immature fibre content	7.6	–	–	12.0
8	Maturity ratio	0.82	–	0.68	0.79
9	Reflectance (R_d)	73.4	–	–	67.8
10	Yellowness (+b)	9.8	–	–	12.6
11	Colour grade	32–1	–	–	34–1
12	Fibre density (g/cc)	1.54	1.459	0.97	0.92–0.95
13	Fibre diameter (µm)	12–23	22	10–21	15–25
14	Moisture content (%)	7.8	–	10	–
15	Moisture regain (%)	8.5	–	11.1	10.5
16	Fibre friction (µm)	0.33	–	–	0.16

Source [2, 30, 64]

Fig. 11 Stretched arrangement of milkweed fibre in a mature pod (*Source* [2])

The slenderness ratio or aspect ratio (L/D ratio) generally starts at 1:100 for the majority of useful fibers. The L/D ratio of milkweed fibre was found to be around 1:1180, which is well within the definition of a textile fibre [2]. The uniformity ratio (ratio between a 50 % span length and a 2.5 % span length) was slightly lower than cotton due to the greater number of fibres in the short length range indicated by short fibre content. The bundle fibre strength and elongation of milkweed fibre was lower than cotton fibre due to the absence of convolutions or crimp-like structures. The maturity ratio of milkweed fibre was slightly lower than cotton fibre as reflected in the higher IFC values. This could be due to the hollow nature of milkweed fibre. The highly stretched arrangement of fibres in the milkweed pod and the fact that these pods were elongated compared with circular cotton bolls might have resulted in the crimpless structure shown in Fig. 11.

6.1 Tensile Properties of Milkweed Fibre

The tensile properties of milkweed (single fibres) are given in Table 4.

Table 4 clearly shows that cotton fibres have higher tenacity and elongation values than milkweed seed fibres but lower values than milkweed stem fibres. Presence of a very thin single wall and lower crystallinity percentage as a result of the lower content of cellulose components in milkweed seed fibre are considered the main reasons for its lower tensile strength. The initial modulus of raw milkweed

Table 4 Tensile properties of cotton and milkweed fibre

Fiber property	Cotton	Milkweed stem fibre [56]	Milkweed seed fibre [2]
Fibre (D)	3–8	104 ± 17	1.05
Tenacity (g/D)	2.7–3.5	3.5 ± 2	3.73
Breaking elongation (%)	6–9	4.7 ± 3.5	3.05
Initial modulus (gf/D)	55–90	122 ± 68	210.89

Source [2, 56]

Fig. 12 Stress–strain behaviour of milkweed fibre (Reproduced with permission from [63])

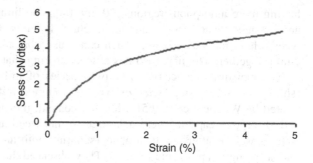

fibre is significantly higher than that of cotton fibre, which could be due to the low breaking elongation value of milkweed compared with cotton.

The general stress–strain behaviour of milkweed fibres analysed by Gharehaghaji and Davoodiis [63] and SEM images of fractured fibres are shown in Figs. 12 and 13, respectively. Figure 12 clearly shows that the circular cross-section of tensile loading has been deformed after loading to a rectangular shape with curved corners which could be due to the thinner wall structure. The fracture study revealed that the fracture behaviour of milkweed fibres is granular, which could be due to their highly oriented molecular arrangement, rather than ductile, as no sign of the latter was observed during tensile fracture of these fibres.

6.2 Moisture Regain

Milkweed fibre exhibits higher moisture regain and moisture content than cotton fibre and hence could be considered a good alternative to commonly used materials for applications which require high moisture absorption. The moisture regain of milkweed fibre has been found to be higher than cotton as a result of the fibre

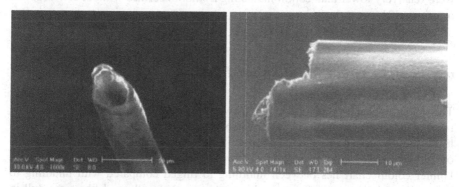

Fig. 13 Fracture pattern of milkweed fibre after tensile loading (Reproduced with permission from [63])

having more amorphous regions [30, 49, 64]. The limited uses of milkweed fibre are strongly related to its unique structural features. One disadvantage is milkweed's high moisture regain, which can cause fibre masses to become damp and clump together. The fibre also has a waxy coating that makes it hydrophobic.

The moisture characteristics of two species of milkweed—*Asclepias speciosa* (showy milkweed) and *Asclepias syriaca* (common milkweed)—have been investigated by Woeppel et al. [65]. Moisture content, moisture regain, absorbency rate and absorptive capacity were measured in milkweed and cotton fibres in the raw state as well as after two treatments (scouring with an aqueous detergent solution and stripping in an organic solvent). They observed that *A. speciosa* and *A. syriaca* fibres were similar in moisture characteristics, indicating that fibres from these two species of milkweed should perform similarly in absorbent materials. Generally, scoured or stripped milkweed fibres exhibited superior moisture content, moisture regain and absorptive capacity than scoured cotton fibers, but could not match the absorbency rate of scoured cotton.

6.3 Fibre Density

The density of raw milkweed fibre was found to be in the range 0.92–0.95 g/cm^3 when the fibre was put over the gradient column without cutting and squeezing and around 1.48 g/cm^3 when the fibre was cut and squeezed to remove air pockets to ascertain the density of the wall without considering hollowness [2].

6.4 Fibre Friction

The fiber-to-fiber friction coefficient was measured on a fibre friction tester. It gave values of 0.33 for cotton fibre and 0.16 for milkweed fibre, respectively. The values were relatively lower than cotton indicating a smooth surface without convolutions or crimps [2].

6.5 Colour Grade of Milkweed Fibre

The colour appearance and grade of milkweed fibre were analyzed by high-volume instrument (HVI) testing. Table 3 shows that the R_d (reflectance) value of milkweed fibre is less than that of cotton and the +b (yellowness) values are higher than those of cotton [2]. The colour grade of milkweed fibre observed from the Nickerson–Hunter colour chart lies between middling to tinged compared with middling to light spotted in the case of cotton, demonstrating that milkweed fibres are dull in white and rich in yellow as a result of the high amount of lignin in the fibres.

6.6 Antibacterial Properties of Milkweed Fibre

Qualitative analysis of antibacterial activity of raw milkweed fibre indicates the presence of a clear zone of inhibition 32–39 mm in diameter against *Staphylococcus aureus* (gram positive) and 21–24 mm against *Escherichia coli* (gram negative), as shown in Fig. 14. Raw milkweed fibre showed higher antibacterial activity against *S. aureus* than *E. coli*. Further, alkali-treated and dyed milkweed fibre does not show any antibacterial activity.

Table 5 and Fig. 15 clearly demonstrate that raw milkweed fibre shows a high percentage of bacterial reduction against *Staphylococcus aureus* (about 93.7 %), but a low percentage against *Escherichia coli* (about 16.77 %).

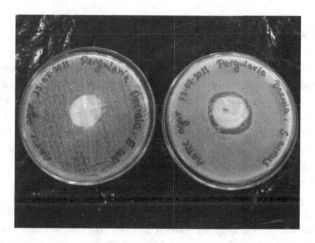

Fig. 14 Antibacterial activity of raw milkweed fibre against *Staphylococcus aureus* and *Escherichia coli*

Table 5 Percentage bacterial reduction against *Staphylococcus aureus* and *Escherichia coli* as a result of quantitative analysis

Serial no.	Dilution rate	*Staphylococcus aureus*		*Escherichia coli*	
		0 h contact time	24 h contact time	0 h contact time	24 h contact time
1	10^{-1}	TNTC	TNTC	TNTC	TNTC
2	10^{-2}	TNTC	156	TNTC	TNTC
3	10^{-3}	248	92	TNTC	TNTC
4	10^{-4}	206	TFTC	298	248
5	10^{-5}	124	TFTC	242	196
6	10^{-6}	TFTC	TFTC	196	144

TNTC, too numerous to count; TFTC, too few to coun

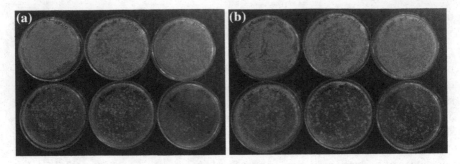

Fig. 15 Percentage bacterial reduction against **a** *Staphylococcus aureus* and **b** *Escherichia coli* as a result of quantitative analysis

7 Spinning of Milkweed Fibre

The main constraint to spinnability of milkweed fibre is its smooth rod-like structure—absent convolutions or crimp-like structures—which makes it more brittle and stiffer with low elongation properties. As a result of its low elongation, it is thought that it could get damaged during mechanical opening (by means of sharp metallic wires) and cleaning in blow room/carding room machines. Further, practical difficulties have been found to be high while processing milkweed fibre—such as web falling, fibre loss, sliver breakage, roller lapping and end breakages in downstream processes [66, 67].

Gharehaghaji and Davoodi [63] analysed the behaviour of milkweed fibre during carding by observing fibre rupture and fibre loss. SEM images of various forms of milkweed fibre during carding are shown in Fig. 16. The sawtooth wire used for carding is the main reason for milkweed fibre damage.

Louis and Kottes [30] effectively produced yarn and fabric in different proportions by blending milkweed fibre with cotton fibre. They found that the addition of milkweed generated more waste when processing blends of cotton/milkweed than 100 % cotton, apparently due to milkweed's lack of cohesiveness. Fibre loss in carding while processing 100 % cotton is around 6.8 %; for 67 % cotton and 33 % milkweed it is about 17.98 %. Further, in addition to major mechanical fibre loss the breaking strength of yarns and fabrics made from cotton/milkweed blends were much lower than 100 % cotton yarns and fabrics.

Yarn production from the stem of *Calotropis procera* (a milkweed species called "aak") was attempted by Varshney and Bhoi [50], who observed that the fineness and strength of milkweed bast fibre were comparable with cotton. Nevertheless, small staple length, high percentage of short fibres and lack of convolutions presented more problems during spinning. Further, they stated that yarn produced from cotton and aak blended 1:1 was inferior to cotton regarding strength, fineness and evenness. This indicated that aak cloth has high tensile and abrasion strength and more weight per square metre than cotton cloth.

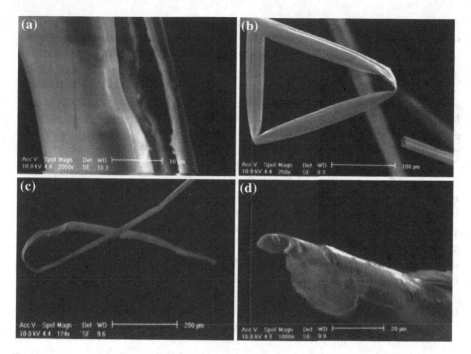

Fig. 16 Various forms of milkweed fibre in card web: **a** fractures in outer surface; **b** bending of fibres; **c** trailing hook of fibre; **d** transverse crack and crack growth (Reproduced with permission from [63])

Andrews et al. [51] found a 38 % increase in strength for 75/25 cotton/milkweed–blended yarn fabric after ammonia mercerization and 66 % after sodium hydroxide mercerization. The corresponding increase for 67/33 cotton/milkweed after ammonia mercerization was 63 and 40 % after sodium hydroxide mercerization, respectively, and the strength decreased after cross-linking. Further, they stated that moisture retention in the cotton/milkweed blend increased after swelling treatments. The effect of swelling treatments overshadowed the decrease in crystalline cellulose by replacing cotton with milkweed.

Drean et al. [47] studied the transverse dimensions (wall thickness and linear density) of milkweed fibre. They analysed the capacity of milkweed's physical fibre properties to process different yarns blended with cotton/milkweed fibre as well as the effect of high milkweed content on transforming these yarns into plain weave fabrics. They found that processing difficulty increases as the proportion of milkweed increases in the blend. It appears to be impossible to spin a pure milkweed fibre yarn using a classical ring-spinning process.

Wazir and Shah [68] analysed different extraction methods of fibre from the bast of *Calotropis procera*. They found that tenacity decreased with increasing relative humidity. Bast fibres of *C. procera* were separated by retting; stem retting for eight days produced the greatest quantity of fibres. Yarn made from a 50/50 blend of cotton and aak was inferior to cotton regarding strength, fineness and evenness. The strength of blended yarn was 50 % that of cotton yarn. Weight per square metre of blended yarn cloth was 2.5 times that of cotton cloth as a result of coarser count and greater thickness.

Sakthivel et al. [64] investigated the spinning of mudar (*Calotropis procera*) fibres in the cotton-spinning system. They found that 100 % mudar yarn could not be produced in the system. The smooth, straight fibre contour of milkweed makes it difficult to spin into yarn. The lack of cohesiveness of milkweed fibre causes extreme difficulties in textile processing.

Karthik [2] investigated the relationship between the yarn structure and yarn properties of cotton/milkweed blends in ring- and rotor-spinning systems by analysing fibre migration and the packing density of yarns. He found that decreased tensile properties of cotton/milkweed–blended yarns compared with 100 % cotton yarns was the result of the lower migration factor of milkweed fibre compared with cotton fibre in cotton/milkweed–blended yarns. The radial packing curves of ring- and rotor-spun cotton/milkweed–blended yarns are shown in Figs. 17 and 18, respectively.

Figures 17 and 18 show that the radial packing density of cotton/milkweed–blended yarn decreases from core to sheath of the yarn's structure. Further, the overall packing density of cotton/milkweed–blended yarn decreases with increases

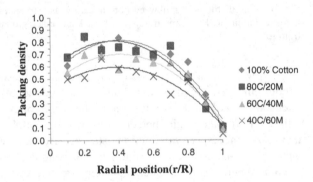

Fig. 17 Radial packing density curves of ring-spun blended yarn (*Source* [2])

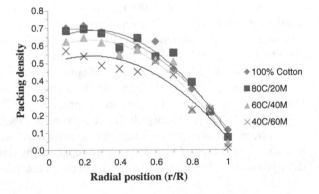

Fig. 18 Radial packing density curves of rotor-spun blended yarn (*Source* [2])

in milkweed fibre percentage, which could also be the result of poor fibre migration. Hence, poor fibre migration and packing density in cotton/milkweed blends are responsible for inferior yarn properties.

7.1 Surface Modification of Milkweed Fibre

The results of numerous researchers demonstrate that spinning 100 % milkweed fibre is practically impossible owing to its inherent characteristics. A great deal of research has been carried out to improve its spinnability by modifying its surface characteristics. The following methods have been tried to improve the same:

- Blending of milkweed fibres with natural and synthetic fibres to improve fibre-to-fibre friction and draftability (Louis and Andrews 1987) [37, 47, 64, 67].
- Surface modification of milkweed fibre by alkali treatment [64, 67].
- Surface modification of milkweed fibre by dyeing [2].
- Surface modification of milkweed fibre by binders/spinning oil [69].
- Surface modification of milkweed fibre by plasma treatment [63].

Sakthivel et al. [2, 64] improved the spinnability of milkweed fibre by delignification (an alkali process that removes lignin). Delignification of milkweed fibre involves treatment with 5 % sodium hydroxide for 30 min at room temperature, washing with acid and water to remove any traces of alkali and drying. SEM and optical microscope images of milkweed fibres after alkali treatment are shown in Figs. 19 and 20, respectively. After alkali treatment the fibre—being hollow—collapses owing to partial removal of hemicellulose and lignin and forms convolutions like cotton (as shown in the figures), which increases milkweed fibre friction to 0.28 µ compared with raw milkweed fibre of 0.16 µ. The tenacity and elongation of alkali-treated milkweed fibre are slightly higher than those of raw fibre. This may be due to rearrangement of molecular chains and formation of convolutions after alkali treatment [49].

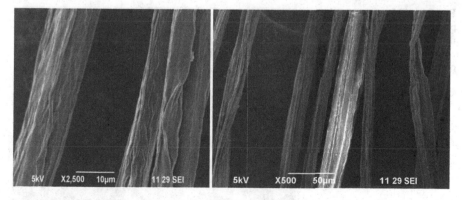

Fig. 19 SEM images of alkali-treated milkweed fibre (*Source* [2])

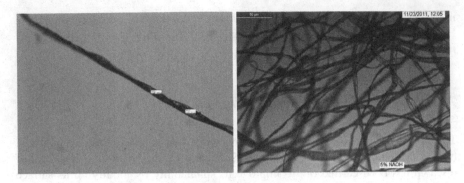

Fig. 20 Optical microscope images of alkali-treated milkweed fibre

Bahreini and Kiumarsi [70] investigated the spinnability and dyeing behaviour of the seed fibre of stabraq (milkweed). They found that raw milkweed fibre and blended cotton/stabraq 75/25 become spinnable after chemical treatment such as scouring and bleaching. Karthik [2] found a 10–15 % increase in yarn strength of cotton/milkweed–blended ring and rotor yarn when cotton is blended with alkali-treated milkweed fibre.

Gharehaghaji and Davoodi [63] modified the surface of milkweed fibre by cold plasma treatment in an attempt to impart surface roughness and thereby increase fibre friction leading to decreased fibre loss and improved spinnability. The treatment involved using the physical vapour deposition method to subject the fibre to nitrogen for 15 min at adjusted power and vacuum pressures. The change in the surface of milkweed fibre after plasma treatment is shown in Fig. 21. Plasma treatment of milkweed fibre clearly leads to fibre surface corrosion. Further, fibre strength and elongation was found to be significantly reduced after plasma treatment; hence it was concluded that milkweed fibre surface modification by plasma treatment is not a suitable way to increase fibre cohesion.

Fig. 21 SEM image of cold plasma–treated milkweed fibre showing surface corrosion (Reproduced with permission from [63])

The main limitations to modifying the surface of milkweed fibre by means of alkali treatment are fibre collapse and fibre matting after treatment. Loss of the hollow nature of the fibre as a result of collapse would mean that desired thermal and sound insulation properties—such as those of raw milkweed fibre—would be unavailable. The matting or stickiness of fibres after alkali treatment would need pre-opening of fibres before carding a second time, which could damage the fibres. To overcome the above problem, [69] tried another method of surface modification—application of wáter-soluble binder—before processing. They found that interfibre friction increases with increased application of binder and concluded that 15 % binder concentration is the optimum for surface modification.

8 Fabric Properties of Milkweed Fabric

Louis and Andrews (1987) studied the fabric properties of cotton/milkweed–blended yarns and the strength retention of fabrics after dimethylol dihydrox-yethyleneurea (DMDHEU) treatment. Untreated cotton/milkweed–blended fabric strengths were much lower than 100 % cotton yarn fabric. They found that the fabric strength retention of cotton/milkweed blends after DMDHEU finishing was greater than that of cotton fabrics. The higher percentage of milkweed in the blend led to higher fabric-breaking strength. The strength responses of chemically modified milkweed blends and the morphological responses of milkweed fibres to swelling might have been due to a combination of the presence of lignin in milkweed and a less ordered cellulosic structure. Varshney and Bhoi [50] investigated the quality of yarns and fabrics produced from milkweed stem fibre. They found that milkweed cloth had high tensile and abrasion strength and more weight per square metre than cotton cloth.

Karthik [2] investigated the comfort poperties of cotton/milkweed–blended rotor yarn fabric—the results are shown in Table 6. The table shows that the air permeability of cotton/milkweed 60/40–blended fabric is significantly lower than 100 % cotton fabric. The poor packing density of cotton/milkweed 60/40–blended yarn along with higher yarn hairiness compared with 100 % cotton yarn resulted in reduced thread spacing in the fabric, which is the reason for the poor air permeability in cotton/milkweed–blended fabric.

Table 6 Comfort properties of 100 % cotton and cotton/milkweed 60/40–blended fabric

Properties	100 % cotton	Cotton/Milkweed 60/40
Air permeability ($cm^3 \cdot cm^2/s$)	113.227	106.309
Moisture vapour permeability ($g/m^2/day$)	1230.408	1217.748
Thermal conductivity (W/mK)	0.0434	0.0278
Wicking (cm)	9.6	13.5

Source [2]

Table 6 clearly shows that the water vapour permeability of cotton/milkweed–blended fabric is significantly lower than 100 % cotton fabric. The increased diameter of cotton/milkweed–blended yarn resulted in reduced interyarn spaces in the fabric and, hence, lower water vapour transmission than 100 % cotton fabrics. The thermal conductivity of cotton/milkweed 60/40–blended fabric is significantly lower than 100 % cotton fabric, which indicates better thermal resistivity or thermal insulation properties than cotton fabrics. The hollow nature of milkweed fibre and larger interfibre gap in the yarn owing to the poor packing density helps air pockets to form both within and between fibres, respectively, resulting in the lower thermal conductivity of blended yarn fabric. The wicking heights of 100 % cotton and cotton/milkweed 60/40–blended yarn fabric after 15 min are 9.6 and 13.5 cm, respectively. The wicking properties of cotton/milkweed 60/40–blended fabric are significantly higher than 100 % cotton fabric. The higher wicking height of cotton/milkweed 60/40–blended fabric showed better capillary effects owing to the hollow nature of milkweed fibre and reduced packing of fibres inside the yarn.

9 Applications of Milkweed Fibre

Milkweed fibre used to be employed for food and medicinal purposes. The floss found in milkweed pods is shaped like a tube and consists of a hollow cellulosic fibre with thin walls that account for little more than 10 % of its total diameter. This provides important advantages regarding absorption. The fibre not only traps air on its surface but also inside. From the acoustic standpoint, these microtubes of air act as barriers against sound. From the thermal standpoint, maintaining empty micro-volumes of almost still air reduces heat exchanges to such an extent that goose down performance—the reference in insulation—is reached.

Water repellence is very high in the best hydrophobic fibres. For example, lotus leaves are not only water repellent they are also very difficult to wet—a property known as the "lotus effect". Lotus leaves allow water to flow without getting wet. They are similar to milkweed fibre in that the angle of contact between the surface and droplets is very high, stopping water from clinging. The low-density, hydrophobic and hollow nature of milkweed fibre has led to active investigation of potential applications in clothing, oil sorption and technical textile manufacture.

9.1 Application of Milkweed Fibre for Clothing Manufacture

The most encouraging application of milkweed fiber is as a loose fill material for jackets and continental quilts. Milkweed seed fibre is often blended with down because the blend demonstrates insulative properties very similar to down. Down is superior to milkweed floss in loftiness and compressibility, which influence product performance, but the properties of milkweed floss can be enhanced by blending with down [43].

Cotton/milkweed–blended yarn fabric has been analysed for its comfort properties by a number of researchers. Karthik [2] investigated the comfort properties of cotton/milkweed blends in ring- and rotor-spinning systems and found that milkweed-blended fabrics had better drapeability and were lighter than 100 % cotton fabrics. Further, milkweed-blended fabrics had better wicking properties and thermal insulation but poorer air and water vapor permeability than 100 % cotton yarn fabric. These results were confirmed by Bakhtiari et al. [71], who compared the comfort properties of milkweed-knitted fabric with those of fabric produced from cotton/milkweed–spun yarns. The comfort properties of milkweed fibre make it suitable to be used as dewaxed fibres with more absorbency than similar cotton products for manufacturing different textiles such as towelling, nappies, bandaging gauze, sanitary napkins, tampons, cosmetic wipes and many other personal care products (Louis and Andrews 1987).

9.2 Application of Milkweed Fibre for Composites

The low density of milkweed fibre (0.9 g/cc) allows the integration of more fibre per unit weight during composite manufacturing; this resulted in manufacturing lightweight composites with better properties [72]. Lightweight composites are finding applications in the automotive industry where weight is critical in light of weight restrictions and where the inherent voids of these composites can enhance sound absorption. Reddy and Yang [72] produced lightweight composites using milkweed fibre as reinforcement and polypropylene (PP) as matrix at different weight ratios and compared them with kenaf-reinforced composites. They found that PP composites reinforced with milkweed fibre had much better flexural and tensile properties than similar PP composites reinforced with kenaf fibre, but they were stiffer than kenaf fibre–reinforced PP composites.

Ashori and Bahreini [29] analysed the potential applications of milkweed stem and seed fibre–reinforced composites and concluded that both types of fibres had good potential for replacing or supplementing other fibrous raw materials as reinforcing agents. Nourbakhsh et al. [73] investigated the potential of Iranian-grown giant milkweed fibres (GFs) as a reinforcing material for thermoplastic composites with PP as matrix along with a coupling agent. They varied the ratio of reinforcing fibre and coupling agent to study its influence on the mechanical properties of composites. They found that the tensile properties of composites increased with increased coupling agent concentration, but that Izod impact strength decreased significantly with increased milkweed fibre concentration—Izod impact testing is named after the English engineer Edwin Gilbert Izod (1876–1946).

Similar studies have been conducted by Srinivas and Babu [74] using milkweed fibre as a reinforcing material in the manufacture of composites. It was found that the mechanical properties—such as tensile strength, tensile modulus, flexural strength and impact strength—of composites improved with increased milkweed fibre fraction. The influence of chemical treatment of milkweed fibres on the mechanical properties of milkweed fibre–reinforced composites has also been studied; it was found that chemical treatment of the fibre enhanced the tensile and flexural properties of composites.

The effect of milkweed fibre on cement-reinforced composite structures—used to minimize the microcracks that form on the surface of composites—has been investigated by Merati [75]. It was found that milkweed fibre was better able to bridge microcracks within the concrete matrix even when mixed at very low percentages during composite manufacture.

9.3 Application of Milkweed Fibre for Oil Spill Clean-up

In March 1989 the oil tanker *Exxon Valdez* spilled 11.2 million gallons of crude oil into the coastal waters of Prince William Sound, causing severe environmental pollution. The destruction of oil storage tanks in Kuwait during the war in 1991 spilled several hundred million gallons of oil into the sea. In addition, oil spills onto land are also common occurrences even though most spills are usually small. The presence of sorbent materials in an oil spill area facilitates a change of phase from liquid to semisolid. Once this change is achieved, the removal of the oil by removal of the sorbent structure is not difficult. While hydrophobicity and oleophilicity are primary determinants of successful sorbents, other important factors include retention of oil over time, recovery of oil from sorbents, amount of oil sorbed per unit weight of sorbent and reusability and biodegradability of sorbent. Of the synthetic products, PP and polyurethane foam are the most widely used sorbents in oil spill clean-up because of their highly oleophilic and hydrophobic properties. A disadvantage of these materials is that they degrade very slowly compared with mineral or vegetable products.

Milkweed seed fibre is expected to have high oil sorption capacity owing to its high wax content—which makes it hydrophobic—and its hollow structure. Choi and Cloud [76] investigated the oil sorption behaviour of cotton, milkweed and kenaf fibre and compared it with PP fibre. It was concluded that milkweed floss showed the highest oil sorption capacity followed by cotton fibre. Further, it was found that alkali scouring of fibres significantly reduced the oil sorption capacity of natural fibres. Choi and Moreau [77] investigated the oil sorption capacities and sorption mechanisms of natural and synthetic fibres and found that the such capacities were higher in natural fibres than synthetic fibers (over 30 g oil/g of fibre). Further, they analysed the sorption mechanism using an environmental SEM, which revealed that oil deposits disappeared from fibre surfaces after a certain time interval in milkweed, kapok and cotton suggesting that the sorption of oil in these fibres occurred through capillary action, probably owing to their hollow lumens.

Fig. 22 Oil sorption capacity
of polypropylene and
milkweed fibres (Reproduced
with permission from [83])

Rengasamy et al. [78] investigated the oil sorption behaviour of fibre assemblies of PP, kapok and milkweed as well as blends of these fibres in filled fibre assemblies and non-woven structure assemblies. They found that the porosity of assemblies played an important role in oil sorption behaviour because fibres were packed closer together and porosity decreased allowing less oil to be absorbed. As porosity decreases, milkweed seed fibre excretes more air bubbles. This happens when the lumen collapses because of pressure. They also found that PP assemblies had higher oil sorption capacity (g/g) followed by kapok and milkweed fibres with porosity <0.98. However, at higher porosities (>0.98), PP-filled fibre assemblies showed poor sorption capacity owing to large interfibre pores, whereas kapok and milkweed fibres with intrafibre porosity showed better oil sorption. They concluded that better oil sorption behaviour, slow degradation of kapok and milkweed fibres due to low cellulose content and biodegradability are the main features these fibres bring to oil spill clean-up. The oil sorption behaviour of PP and milkweed fibres are compared in Fig. 22.

Milkweed seed fibre can be used to absorb crude oil. Using a coarse-mesh plastic net, fibres can be contained and removed. Using 0.027 g of fibres in a 10 × 10-cm pad, 10 mL of engine oil with a specific gravity of 0.75 g/mL can be absorbed.

9.4 Application of Milkweed Fibre for Insulation

A study of the insulation properties of various materials found that milkweed fibre owing to its hollow structure would possess significantly higher sound absorbency. Crews et al. [43] investigated the potential application of milkweed fibre as insulation material. They analysed seven identical jackets filled with various materials on a per unit weight basis and measured the thermal insulation, thickness, compression, resiliency and hand. They concluded that milkweed floss blended with down had insulative properties similar to down.

Hassanzadeh et al. [79] studied the acoustic properties of needled non-woven fabrics produced from naturally grown hollow estabragh fibres and PP using sound impedance tube and optimized parameters using the Taguchi method. They concluded that blends of milkweed and PP fibres in the form of lightly needled non-woven fabrics could potentially be utilized for sound pollution control. This can mainly be attributed to the hollow structure of milkweed fibre in the blend. Hasani et al. [80] investigated the noise absorption coefficient (NAC) of milkweed and hollow polyester fibre blends by varying the blend ratio, mass per unit area, punching density and frequency of incident sound. They found that the NAC value increased with increase in the ratio of milkweed fibre in the blend owing to its hollow nature. The whole-diameter-to-fibre-diameter ratio of milkweed fibre is much higher than that of hollow polyester fibre. Hence, a higher milkweed ratio in the blend could increase the surface area of non-woven fabrics. This resulted in more frictional losses of sound energy in the non-woven samples leading to higher NAC values.

9.5 Potential Application Areas of Milkweed Fibre

9.5.1 Superabsorbent Fibre

Milkweed fibre has the capacity to absorb 75 g/g of saline solution after treatment with a surfactant. Though wood pulp with the same amount of fibre could absorb about the same volume of liquid, owing to the lesser weight of milkweed (one-fifth that of wood pulp) there are more milkweed fibres per unit weight than pulp [55].

9.5.2 Fluid Carrier Fibre

Milkweed fibre readily absorbs fluids, but rigidly retains only about 5 % by weight of these fluids. This small percentage of absorbed fluids appears to enter the lumen of floss fibre and is strongly retained by surface tension [55]. This feature of milkweed fibre could be used as a chemical carrier in non-woven products to produce a biodegradable package.

9.5.3 Self-bonding Fibre

Milkweed fibre has a thin wax coat which allows weak bonding to adjacent fibres when exposed to heat and low pressure or to solvents. The outer cellulosic layers of these fibres are very thin. This characteristic allows easy bonding of inner ligneous material to adjacent fibres when subjected to more heat and greater pressure [55].

Fig. 23 Bacterial growth
inside milkweed fibre
(Reproduced with permission
from [81])

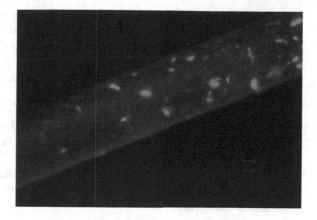

9.5.4 Bio-active Fabrics

Research is under way to create fabric-based bio-reactors in which colonies of mammalian cells or bacteria can live and function for extended periods of time. The inside of milkweed's hollow fibres has been utilized to grow bacteria, as shown in Fig. 23 [81].

10 Sustainability and Challenges

Milkweed plants are presently taken from the wild. Yield can be improved by cultivating milkweed rather than taking it from the wild. Cotton harvesting has suffered severe setbacks owing to its vulnerability to pests. Huge amounts of pesticides have to be applied with deleterious consequences for the health of people and the quality of the environment. Fertilizer is also necessary, resulting in much higher costs and less net income for cotton farmers. Furthermore, cotton has to be grown on fairly good land; this is becoming scarcer owing to the ever increasing pressure for land to grow food crops to satisfy the needs of a rapidly increasing world population. Marginal land left over from shifting cultivation, mining and other human activities, despite being abundant worldwide, cannot be utilized for growing cotton unless the high cost of land improvement, irrigation and fertilizers can be borne by the farmers themselves.

As every gardener or farmer is aware, weeds grow quite happily without any help whatsoever. Hence, the surprise when certain weeds with properties so useful they deserve to be cultivated often prove very difficult to cultivate. It is important to keep in mind that weed fibres that have been investigated and found to have potential to supplement and substitute other natural fibres for the production of fabric or any other application have a number of things going for them [82]:

- waste land or unutilized land can be used for their cultivation;
- the substantial work involved in cultivation such as irrigation and fertilizer use could be minimized significantly resulting in cost reduction and higher income for farmers; and
- using fewer pesticide lessens health risks and damage to the environment.

The cultivation of milkweed offers several environmental benefits:

- it does not need insecticides, fertilizers or artificial watering;
- its huge root system reinstates organic matter in the soil and maximizes microbiological activity;
- it supports pollinating insects such as bees and butterflies; and
- no chemical treatment is required for its transformation from its status as a weed to a cultivated plant.

Ironically, the principal difficulty in growing milkweeds is they can be swamped by other weeds; this is certainly something to keep in mind when attempting to grow them. Equally ironically, glyphosate herbicide has been used by some growers of milkweeds to control weeds in attempts to establish plants to conserve monarch butterflies (*Danaus plexippus*). Diseases and pests (including monarchs) are also significant problems. As noted later, using pesticides to control milkweed pests could pose a serious problem not only for monarch butterflies, but also for several other species of butterflies.

11 Conclusions and Recommendations

The search continues for the ideal natural fibre—one that is organically cultivated with zero or minimal artificial assistance, ethically manufactured, sustainable, processed without chemical aid, with reusable by-products and completely bio-degradable. The gross properties of numerous varieties of milkweed fibre have been found to be similar but variations can be seen owing to different soil conditiona. The main positive attributes of milkweed fibre are its low density, hollow nature and hydrophobicity which make it suitable for oil sorption and thermal and sound insulation applications. The main limitations to converting milkweed fibre into yarn are its low elongation and brittle nature which makes it impractical to process 100 % milkweed fibre yarns but warrants blending with natural and synthetic fibres.

The Natural Fibers Corporation, Ogallala, Nebraska (USA) is currently cultivating and marketing milkweed floss commercially at a price of $28/kg. As the corporation has a monopoly in this market the prices are relatively high. The prices will come down when more players enter the market. Nurseries are increasingly offering milkweed stock for gardeners and conservationists. Native plant societies also provide seeds for planting. Milkweed could be employed for a number of purposes: landscaping, habitat restoration, erosion control and naturalizing highway areas. Milkweed seed oil is used in the manufacture of cosmetics, industrial specialty chemicals and dietary extracts. Milkweed plants are vital for butterfly conservation and habitat restoration.

Successful commercialization of milkweed as a crop hinges on mechanizing harvesting, handling, drying and floss processing. The yield of milkweed plants can be improved by first selecting superior strains and hybrid seeds and then using appropriate cultivation methods. The successful production of spun yarns from milkweed fibre entails finding suitable chemical treatments that can modify its smooth surface without affecting its hollow nature. Since milkweed fibre is fine and less dense than cotton, ginning machines used for cotton cannot be used for milkweed fibre. Though seeds are removed manually for research work, commercial production will require a ginning machine that takes into account the specific properties of milkweed fibre. Furthermore, beaters and other blow room components as well as wire points and suction systems in carding machines need to be redesigned with the properties of milkweed fibre in mind so that they can be processed for mass production and commercialization. The above points clearly demonstrate that the conversion of milkweed fibre to yarn will likely be a much more difficult task than converting it to web for non-woven or composite applications.

Research work is currently focussed on utilization of milkweed fibre for oil spill clean-up. Various chemical treatments have been tried to make the fibre only absorb oil rather than water. Some research has revealed potential applications for milkweed-blended fabrics as thermal insulation garments; however, practical implications of the same are in question. Milkweed could become a significant industrial crop in the near future—it is already vital to supporting butterflies and preserving their habitats.

References

1. Brundtland GH, Khalid M (1987) Our common future, report of the world commission on environment and development. Oxford University Press, Oxford
2. Karthik T (2014) Studies on the spinnability of milkweed fibre blends and its influence on ring, compact and rotor yarn characteristics. Ph.D. thesis, Anna University
3. Ashori A (2006) Non-wood fibers—a potential source of raw materials in paper making. Polym Plast Technol Eng 45:1133–1136
4. Mauersberger HR (1962) Matthews' textile fibers, 6th edn. Wiley, New York
5. Subramanian K, SenthilKumar P, Jeyapal P, Venkatesh N (2005) Characterization of ligno-cellulosic seed fiber from Wrightia Tinctoria plant for textile applications. Eur Polym J 41:853–861
6. Vijayraaghavan NN, Karthik T (2004) Multi component fiber technology for medical and other filtration applications. Synth Fiber 33(1):5–8
7. Campbell TA, Grasse KA (1986) Effect of stage of development on chemical yields in common Milkweed—Asclepias syriaca. Biomass 9:239–246
8. Karthik T, Gopalakrishnan D (2014) Environmental Analysis of Textile Value Chain: An Overview. In: Muthu SS (ed) Roadmap to sustainable textiles and clothing, Textile Science and Clothing Technology. Springer, Singapore, pp 153–188
9. Mohanty AK, Misra M, Drzal LT (2002) Sustainable biocomposites from renewable resources: opportunities and challenges in the green material world. J Polym Environ 10:9–26

10. Eichhorn SJ, Baillie CA, Zafeiropoulos NE, Mwaikambo LY, Ansell MP, Dufresne A (2001) J Mater Sci 36:2107–2112
11. Karthik T, Rathinamoorthy R (2015) Recycling and reuse of textile effluent sludge. In: Muthu SS (ed) Environmental implications of recycling and recycled products. Environmental footprints and eco-design of products and processes. Springer, Singapore, pp 213–258
12. Rout J, Misra M, Mohanty AK, Nayak SK, Tripathy SS (2003) SEM observations of the fractured surfaces of coir composites. J Reinf Plast Compos 22(12):1083–1100
13. Bledzki AK, Gassan J (1999) Composite reinforced with cellulose based fibers. Prog Polym Sci 24:221–274
14. CIRCOT (1999) An insight into ligno-cellulosic fibres: structure and properties. CIRCOT, Mumbai, pp 89–94
15. Chaitanya MVNL, Dhanabal SP, Rajendran RS (2013) Pharmacodynamic and ethnomedicinal uses of weed speices in nilgiris, Tamilnadu State, India: a review. Afr J Agric Res 8(27): 3505–3527
16. Tuntawiroon N, Samootsakorn P, Theeraraj G (1984) The environmental implications of the use of calotropis gigantea as a textile fabric. Agric Ecosyst Environ 11:203–212
17. Witt MD, Knudsen HD (1993) Milkweed cultivation for floss production. In: Janick J, Simon JE (eds) New crops. Wiley, New York, pp 428–431
18. Muenscher WC (1975) Poisonous plants of the United States (Rev. ed.). Collier MacMillan, New York, pp 195–199
19. Berkman B (1949) Milkweed—a war strategic material and a potential industrial crop for submarginal lands in the United States. Econ Bot 3:223–239
20. Moore RJ (1946) Investigations on rubber bearing plants. IV. Cytogenetic studies in Asclepias Torr. L. Can J Res (Sect C) 24:66–73
21. Stevens OA (1945) Cultivation of milkweed. North Dakota Agric Exp Stat Bull 333–335
22. Gaertner EE (1979) The history and use of milkweed (Asclepias Syriaca L.). Econ Bot 33 (2):119–123
23. Lobstein MB (2013) There and back again: a short taxonomic history of milkweed. http:// vnps.org/princewilliamwildflowersociety/botanizing-with-marion/there-and-back-again-a-short-taxonomic-history-of-milkweed/. Accessed on 13 August 2015
24. Parrotta A (2001) John Healing plants of peninsular India. CABI Publishing, New York
25. Lichtenstein MHC (1839) Zur Erledigung Dee Frage Uber Die Benutzung Der Syrischen seidenpflanze (asclepias syriacA). Ver. z. Beford, Gartenbaues, K. Preuss. Staaten Vernandl. 14(2):218–220
26. Ulbricht H (1940) Die bastfasern von asclepias syriaca L. Forscbung 14(4):232–237
27. Knudsen HD, Sayler RY (1993) Milkweed: the worth of a weed. In: New crops, new uses, new markets 1992 ycar book of a agriculture, US Dept of Agriculture, Washington, pp 118–123
28. Nehring J (2013) The potential of Milkweed Floss as a natural fiber in the textile industry. J Undergrad Res 13:64–68
29. Ashori A, Bahreini Z (2009) Evaluation of Calotropis gigantean as a promising raw material for fiber-reinforced composite. J Compos Mater 43:1297–1304
30. Louis GL, Kottes BA (1987) Cotton/milkweed blends: a novel textile product. Text Res J 57(6):339–345
31. Adams RP, Balandrin MF, Martineau JR (1984) The showy milkweed, Asclepias speciosa: a potential new semi-arid land crop for energy and chemicals. Biomass 4(2):81–104
32. Borders B, Lee-Mäder E (2014) Milkweeds—a conservation practitioner's guide. The Xerces Society for Invertebrate Conservation, Portland
33. Phippen WB (2007) Production variables affecting follicle and biomass development in common milkweed. In: Janick J, Whipkey A (eds) Issues in new crops and new uses. ASHS Press, Alexandria, pp 82–87
34. Roşu A, Danaila-Guidea S, Dobrinoiu R, Toma F, Rosu DT, Sava N, Manolache C (2011) Asclepias syriaca L.—an underexploited industrial crop for energy and chemical feedstock. Rom Biotechnol Lett 16:131–138

35. Beckett RE, Stitt RS (1935) The desert milkweed (Asclepias subulata) as a possible source of rubber. Technical Bulletin No. 472, U.S. Department of Agriculture, Washington, D.C, 23 pp
36. Knudsen HD, Zeller RD (1993) The milkweed business. In: Janick J, Simon JE (eds) New crops. Wiley, New York, pp 422–427
37. Crews PC, Rich W (1995) Influence of milkweed fiber length on textile product performance. Cloth Text Res J 13(4):213–219
38. Heise PJ, Vidaver AK (1989) A new xanthomad: casual agent of bacterial blight of milkweed. In: Proceedings of the seventh international conference on Plant pathology pp 547–552
39. Witt MD, Nelson LA (1992) Milkweed as a new cultivated row crop. J Prod Agric 5(1):167171
40. Wilbur HM (1976) Life history evolution in seven milkweeds of the genus Asclepias. J Ecol 64(1):223–240
41. Luna T, Kasten Dumroese R (2013) Monarchs (Danaus plexippus) and milkweeds (Asclepias species). Native Plants 14(1):5–15
42. Stevens M (2003) Common Milkweed. USDA. http://plants.usda.gov/plantguide/pdf/cs_assy.pdf. Accessed on 23 July 2015
43. Crews PC, Sievert SA, Woeppel LT (1991) Evaluation of milkweed floss as an insulative fill material. Text Res J 61:203–210
44. Woodson RE (1954) The North American species of Asclepias L. Ann Mo Bot Gard 41:1–211
45. Anonymous (1944) Milkweed helps solve fiber problem. J Chem Educ 21(2):54–55
46. Evetts LL, Burnside OC (1974) Root distribution and vegetative propagation of Asclepias syriaca L. Weed Res 14:283–288
47. Drean J-YF, Patry JJ, Lombard GF, Weltrowski M (1993) Mechanical characterization and behavior in spinning processing of milkweed fibers. Text Res J 63(8):443–450
48. Karthik T, Murugan R (2013) Milkweed fibres: properties, and potential applications. Melliand Int 3:151–153
49. Karthik T, Murugan R (2013) Characterization and analysis of ligno-cellulosic seed fiber from Pergularia Daemia Plant for textile applications. Fibers Polym 14(3):465–472
50. Varshney AC, Bhoi KL (1987) Some possible industrial properties of calotropis procesra (aak) floss fiber. Bio Waste 22(2):157–161
51. Kottes Andrews BA, Kimmel LB, Bertoniere NR, Hebert JJ (1989) Comparison of the response of cotton and milkweed to selected swelling and crosslinking. Text Res J 59(11):675–679
52. Jones D, Von Bargen KL (1992) Some physical properties of milkweed pods. Trans Am Soc Agric Eng 35(1):243–246
53. Prasad P (2006) Physio-chemical properties of a nonconventional fiber: aak (calotropis procera). J Text Assoc 67(2):63–66
54. Haghighat-Kish H, Shaikhzadeh-Najar S (1998) Structure and properties of a natural cellulosic hollow fiber. Int J Eng 11:101–108
55. Knudsen HD (1990) Milkweed floss fiber for improving nonwoven products. In: TAPPI Proceedings of Nonwovens conference, USA, pp 209–212
56. Reddy N, Yang Y (2009) Extraction and characterization of natural cellulose fibers from common milkweed stems. Polym Eng Sci 49:2212–2217
57. Hessler LE, Merola GV, Berkley EE (1948) Degree of polymerization of cellulose in cotton fibres. Text Res J 18(10):628–634
58. Timell TE, Snyder JL (1955) Molecular properties of milkweed cellulose. Text Res J 25(10):870–874
59. Barth FW, Timell TE (1958) The constitution of a hemicellulose from milkweed (asclepias syriaca) floss. J Am Chem Soc 80(23):6320–6325
60. Goda K, Sreekala MS, Gomes A, Junji TK, Ohji J (2006) Improvement of plant based natural fibres for toughening green composites—effect of load application during mercerization of ramie fibres. Compos Part A 37(12):2213–2220
61. Gu P, Hessley RK, Pan W-P (1992) Thermal characterization analysis of milkweed floss. J Anal Appl Pyrol 24(2):147–161

62. Schwenker RF, Zuccarello RK (1964) Differential thermal analysis of synthetic fibres. J Polym Sci 6(1):1–16
63. Gharehaghaji AA, Hayat-Davoodi S (2008) Mechanical damage to Estabragh fibers in the production of thermobonded layers. J Appl Polym Sci 109:3062–3069
64. Sakthivel JC, Mukhopadhyay S, Palanisamy NK (2005) Some studies on mudar fibers. J Ind Text 35(1):63–76
65. Woeppel LT, Crews PC, Sievert SA (1990) Determining moisture characteristics of milkweed floss. AATCC Book of Papers 172–176
66. Karthik T, Murugan R (2013) Influence of spinning parameters on milkweed/cotton DREF-3 yarn properties. J Text Inst 104:938–949
67. Karthik T, Murugan R (2014) Influence of friction spinning process parameters on spinnability of Pergularia/Cotton-Blended Yarns. J Nat Fibers 11:54–73
68. Wazir TA, Shah SMA (1977) Physical characteristics and utilization of aak (Calotropis procera) fibers. Pak J Forest 27(2):69–80
69. Bahl M, Arora C, Parmar MS, Rao JV (2013) Surface modification of milkweed fibers to manufacture yarns. Text Potpouri Colourage 60:33–35
70. Bahreini Z, Kiumarsi A (2008) A comparative study on the dyeability of stabraq (milkweed) fibers with reactive dyes. Prog Color Color Coat 1(1):19–26
71. Bakhtiari M, Hasani H, Zarrebini M, Hassanzadeh S (2015) Investigation of the thermal comfort properties of knitted fabric produced from Estabragh (Milkweed)/cotton-blended yarns. J Text Inst 106:47–56
72. Reddy N, Yang Y (2010) Non-traditional lightweight polypropylene composites reinforced with milkweed floss. Polym Int 59:884–890
73. Nourbakhsh A, Ashori A, Kouhpayehzadeh M (2008) Giant milkweed (Calotropis persica) fibers—a potential reinforcement agent for thermoplastics composites. J Reinf Plast Compos 28:2143–2149
74. Srinivas CA, Babu GD (2013) Mechanical and machining characteristics of Calotropis Gigentea Fruit fiber reinforced plastics. Int J Eng Res Tech 2:1524–1530
75. Merati AA (2014) Reinforcing of cement composites by Estabragh fibers. J Inst Eng India Ser E 20:1–6
76. Choi H-M, Cloud RM (1992) Natural sorbents in oil spill cleanup. Environ Sci Technol 26:772–776
77. Choi H-m, Moreau JP (1993) Oil sorption behavior of various sorbents studied by sorption capacity measurement and environmental scanning electron microscopy. Microsc Res Tech 25:447–455
78. Rengasamy RS, Das D, Karan CP (2011) Study of oil sorption behavior of filled and structured fiber assemblies made from polypropylene, kapok and milkweed fibers. J Hazard Mater 186:526–532
79. Hassanzadeh S, Hasani H, Zarrebini M (2014) Analysis and prediction of the noise reduction coefficient of lightly-needled Estabragh/polypropylene nonwovens using simplex lattice design. J Text Inst 105:256–263
80. Hasani H, Zarrebini M, Zare M, Hassanzadeh S (2014) Evaluating the acoustic properties of Estabragh (Milkweed)/Hollow-Polyester nonwovens for automotive applications. Text Sci Eng 4:1–6
81. Fowler AJ, Warner SB, Toner M (2002) Development of bio-active fabrics. NTC Project. National Textile Center Research Briefs—Materials Competency
82. Penn State (2015) Introduction to weeds and herbicides. http://extension.psu.edu/pests/weeds/control/introduction-to-weeds-and-herbicides/extension_publication_file. Accessed on 13 Aug 2015
83. McDiarmid M (2014) Milkweed touted as oil-spill super-sucker—with butterfly benefits. http://www.cbc.ca/news/politics/milkweed-touted-as-oil-spill-super-sucker-with-butterfly-benefits-1.2856029. Accessed on 28 July 2015

Printed in the United States
By Bookmasters